Embracing Complexity in Design

This page intentionally left blank

Embracing Complexity in Design

Edited by Katerina Alexiou,
Jeffrey Johnson and
Theodore Zamenopoulos

Routledge
Taylor & Francis Group
LONDON AND NEW YORK

First published 2010
by Routledge
2 Park Square, Milton Park, Abingdon, Oxfordshire OX14 4RN

Simultaneously published in the USA and Canada
by Routledge
711 Third Avenue, New York, NY 10017

Routledge is an imprint of the Taylor and Francis Group, an informa business

First issued in paperback 2015

Typeset in Univers by Graphicraft Limited, Hong Kong

British Library Cataloguing in Publication Data
A catalogue record for this book is available from the British Library

Library of Congress Cataloging in Publication Data
Embracing complexity in design / edited by Katerina Alexiou, Jeffrey Johnson &
Theodore Zamenopoulos.
p. cm.
Includes bibliographical references and index.
1. Design. 2. Complexity (Philosophy) I. Johnson, J. (Jeffrey) II. Alexiou, Katerina, 1973– III.
Zamenopoulos, Theodore, 1973–
NK1520.E43 2010
701'.8–dc22
2009008956

ISBN 978-0-415-49700-8 (hbk)
ISBN 978-1-138-96862-2 (pbk)
ISBN 978-0-203-87139-3 (ebk)

Contents

Tables

Figures

Contributors

Katerina Alexiou is RCUK Research Fellow in Design at the Open University. Her research focuses on the area of design theory and methods, including design cognition (from a neurological and computational perspective), collaborative design, learning, creativity, and social aspects of design. She is also interested in investigating how complexity theories and methods can be applied to the understanding and support of design processes and design thinking.

Leonard R. Bachman is the founder and co-director of the Building Performance Laboratory at the University of Houston, College of Architecture, where he has taught environmental systems, systems integration, and research methods since 1979. His publications include the books *Integrated Buildings* (2003) and *Spreadsheets for Architects* (1998, with David Thaddeus).

Michael Batty is Bartlett Professor of Planning and Director of the Centre for Advanced Spatial Analysis in University College London. He has degrees from the University of Manchester (BA) and the University of Wales (PhD), and is a Fellow of the British Academy as well as a Fellow of the Royal Town Planning Institute, the Chartered Institute of Transport and the Royal Society of Arts. His research is in the development of computer-based technologies, specifically graphics-based and mathematical models for cities, and he has worked recently on applications of fractal geometry and cellular automata to urban structure. He is the editor of the journal *Environment and Planning B* and was awarded the CBE for 'services to geography' in 2004.

Karen Cham is an artist who has worked with creative technologies since 1987, exploring the poetic potential of technology within a critical context. She is responsible for developing Kingston University's Digital Media Institute across four faculties and leading the convergent postgraduate programme. She supervises research degrees, and her research interests include how media semantics might inform computational aesthetics and design methodologies for complex systems.

John Clarkson is Professor of Engineering Design and Director at Cambridge Engineering Design Centre, University of Cambridge. His research interests are in the general area of engineering design, particularly the development of design methodologies to address specific design issues: for example, process management, change management, healthcare design and inclusive design. As well as publishing over 400 papers, he has written a number of books on medical equipment design and inclusive design.

Claudia Eckert has been a senior lecturer in design at the Open University since 2008. Previously she was Associate Director of the Engineering Design Centre in Cambridge. Her research interests are in engineering change, process planning and modelling, and design creativity; and in analysing the similarities and differences between different design domains.

Jeffrey Johnson is Professor of Complexity Science and Design at the Open University and Head of the Department of Design, Development, Environment and Materials. He is also Rector of the Open University for Complex Systems in Paris. He is a chartered mathematician and engineer with extensive experience of applications of complex systems research to the design and management of social systems. He has directed many academic and commercial projects, and published four books and seventy articles in the area of complex systems. He is President of the Complex Systems Society.

René Keller holds a PhD in Engineering Design from Cambridge University in management of engineering change. His research interests include information visualisation and predicting change risks for complex products. As a researcher at Cambridge University he has worked on a number of projects in collaboration with the defence, aerospace and automotive industries. In late 2008 he joined BP.

Alec Robertson is an independent consultant and entrepreneur with www.4d-dynamics.net, and also an academic at De Montfort University, Leicester. Alec's research interests include dissemination issues of design research, along with 'complexity science and 4D design'. He is a member of the Chartered Society of Designers and a Fellow of the Royal Society of Arts.

George Rzevski is Professor Emeritus, The Open University, Milton Keynes, and Founder, Magenta Corporation, London. His current research interests include managing complexity of business and social systems, design of complex adaptive artefacts and creation of emergent intelligence. George has initiated and provided know-how for the design of several large-scale multi-agent systems which are in commercial use earning an excellent return on investment.

Mick Wallis is Professor of Performance and Culture and Pro-Dean Research in the Faculty of Performance, Visual Arts and Communications at the University of Leeds. His interdisciplinary research ranges from performance history – including recently the use of amateur theatre in interwar rural England – to robotics. Before joining academia at 40, he worked in publishing and advertising production.

Robert Young is Professor of Design Practice, Associate Dean for Research and Consultancy in the School of Design and Director of the Centre for Design Research at Northumbria University. His research interests concern the development of design thinking, practice knowledge and theory, including service design, innovation and reflective practice for designers.

Theodore Zamenopoulos is Lecturer at the Open University, working on projects related to the theme of complexity and design and design education. His research is mainly concerned with the identification of the conditions that determine our capacity to carry out creative design tasks. His work includes theoretical investigations on the mathematical conditions that underlie design cognition, and empirical research on the neurological and social basis of design. He has also worked on computational tools to support design creativity and learning.

Editors' preface

Katerina Alexiou, Jeffrey Johnson and Theodore Zamenopoulos

Design is a quintessentially human activity that involves envisioning the future of our natural and artificial environment, creating innovative products and solutions, and intervening in order to produce desirable effects. As the realisation grows that design activities and artefacts display properties associated with complex adaptive systems, so grows the need to use complexity concepts and methods to understand these properties and inform the design of better artefacts. Complexity science represents an epistemological and methodological shift that promises a holistic approach in the understanding and operational support of design.

We consider that embracing complexity in design is one of the critical issues and challenges of the twenty-first century.

But design is also a major contributor in complexity research. Design research is important not only because scientific investigation involves design processes (for example, designing experiments), but also because design provides an avenue for scientific results, and knowledge to influence the way we construct and change the world around us. The linkage between complexity and design is a new cutting-edge topic which is expected to gain more and more appreciation and recognition. Although both the design and the complexity research communities have an awareness of the importance of the other, there is currently no resource that draws out the important research issues that link the two in a focused way.

This book, written to be accessible to the general reader, reports a set of research topics and questions developed through a series of research workshops organised within the Embracing Complexity in Design project. The project was funded under the Designing for the 21st Century Initiative jointly sponsored by the UK Arts and Humanities Research Council and the Engineering and Physical Science Research Council. The wide range of projects funded under that initiative formed a network supporting intense activity across the many areas of design, as recorded in Inns (2007). Our book links the wider design community with the complex systems research community, providing state-of-the-art developments in the area of complexity and design, and synthesises a variety of different design themes and domains.

The book can be read in two ways. One way is to approach each chapter as a contribution to a particular design domain. Batty (Chapter 1) offers a complexity perspective on urban planning and the development of cities; Bachman (Chapter 2) outlines a view of architecture as a complex system; Eckert, Keller and Clarkson (Chapter 3) discuss complexity in engineering design; Rzevski (Chapter 4) talks about technologies for designing complexity in organisations; Alexiou (Chapter 5) and Zamenopoulos (Chapter 6) focus on design theory; Cham (Chapter 7) and Wallis (Chapter 8) explore complexity in digital and performance art; Young (Chapter 9) considers the domain of service design; Robertson (Chapter 10) reflects on interactive design; and Johnson (Chapter 11) explores the place of design at the heart of the new science of complex systems.

The other way is to explore each chapter as putting forward a different perspective on the relation between complexity and design, and how it can be cultivated. For Batty, complexity is an inherent characteristic of natural and artificial systems. Cities in particular are exemplars of the way order can be created from complex bottom-up processes. His chapter shows how methodologies coming from complexity science can inform our understanding of city generation, and discusses implications for planning and intervention. Bachman is also concerned with the built environment, but his focus is on architecture. He discusses how complexity is part of building design, not only in terms of physical processes, but also in terms of the social and cultural phenomena that envelop both the practice and the experience of architecture. Eckert *et al.* view complexity as a problem that we need to manage and reduce in order to avoid risk and make robust and functional products. This is particularly important in large engineering projects, where small changes may propagate disastrous effects. Rzevski, on the other hand, sees complexity as an opportunity for incorporating adaptive behaviours into products, processes and organisations, and discusses how complexity can be built into multi-agent technology systems. Alexiou uses multi-agent systems notions and methodologies in order to develop a theory about design as a collaborative, distributed process. Both Alexiou and Zamenopoulos focus on methodological and epistemological links between complexity and design. Zamenopoulos takes design as a natural phenomenon that not only cuts across different domains, but also exists at different levels of reality (neurological, social or cultural). His chapter proposes a mathematical theory that explains the ability to carry out design tasks. The chapter by Cham explores the relationship between science and art, with an emphasis on the contribution art can make in the science of complex systems. The main argument here is that art includes theories and practices that can inform complex systems science, particularly contributing to the representation and visualisation of complex systems. Wallis looks at the triangle between design, performance and complexity, and suggests that performance can help embody and ground scientific understanding. Young reviews developments in service design, and suggests that complexity may be a helpful way forward for service design thinking and practice. Robertson offers a view on the relationship between complex systems science, cybernetics and design, and envisages future possibilities for innovation in everyday products, services and systems. Finally, Johnson argues that design as *in vivo* experiment is at the heart of the methodology of the emerging science of complex systems. An unexpected outcome of *embracing complexity in design* is that it is essential to *embrace design in the science of complex systems.*

We believe that this book will be a useful resource for anybody entering this new area of investigation from a different background, and we hope that every reader will find threads of research that can usefully inform his or her own academic or professional practice.

Walton Hall, February 2009

Reference

Inns, T. (2007), *Designing for the 21st Century: Interdisciplinary Questions and Insights*, London: Gower Ashgate.

This page intentionally left blank

1
Generating cities from the bottom up

Using complexity theory for effective design

Michael Batty

1 Introduction

This essay introduces the idea that cities evolve from the bottom up, that patterns emerge as a global order from local decisions. As the future is unknowable, cities must be planned according to this type of complexity. We begin by introducing the idea of hierarchy and modularity as the basis of a generative process that leads to patterns that are self-similar over many scales, patterns that are fractal in their structure. We then present a simple model of growth that generates such patterns based on a trade-off between connecting to the growing structure and seeking as much free space as possible around any location within it. These are the tensions that exist in real cities through the quest by individuals to agglomerate. Once we have sketched the model, we demonstrate how this generative process can be simulated using cellular automata, allowing us to think of the logic in terms of rules that represent how existing development takes place. By changing the rules, we can introduce optimality into the process, developing the logic to enable certain goals to be pursued. We conclude with various demonstrations of how idealised plans might be conceived within this complexity which is generated from the bottom up.

2 A new paradigm for city planning

Cities are built from the bottom up. They are the product of millions of individual decisions on many spatial scales and over different time-intervals, affecting both the functioning and the form of the city with respect to how it is structured and how it evolves. It is impossible to conceive of any organisation that can control such complexity, and thus the very question of the extent to which the city might be 'planned' is thrown into a new light. Throughout history, plans for cities have been proposed as a top-down response to perceived problems and the realisation of ideals, but there are few examples where control has been sufficiently strict to enable their complete implementation. Most development in cities occurs without any central planning yet cities continue to function, often quite effectively, without any top-down control. Cities as part of societies and economies not only hold together without any top-down control but actually evolve their own coordination from the bottom up, their order emerging from these millions of relatively uncoordinated decisions which express Adam Smith's characterisation of the economy as being managed by an 'invisible hand'.

Fifty years ago, cities were first considered to be systems whose functioning was based on many interacting parts and whose form is manifested in a relatively coordinated hierarchy of these parts (or subsystems). Yet systems in these terms were conceived of as being centrally controlled. As the paradigm developed, there was a subtle shift to the notion that the order in many systems and their resulting hierarchies emerged from the way their parts or elements interacted from the bottom up rather than from any blueprint imposed from the top down. The complexity sciences developed to refresh this systems paradigm, with the focus changing from an analogy between cities as machines to one based on evolving biologies, whose form was the resultant of subtle and continuous changes in their genetic composition at the level of their most basic component parts. This shift in thinking is wider than cities *per se*. It is from thinking of the world in terms of its physics to one based on its biology, from top down to bottom up, from centralised to decentralised action, and from planned forms to those that evolve organically.

In this chapter, I shall argue that a new paradigm for planning cities is required which takes account of how they are built, which is largely but not exclusively from the bottom up. It draws on recent developments in complexity theory which, in terms of city planning, draws on the traditions first suggested by Patrick Geddes (1915) in his book *Cities in Evolution* at the beginning of the last century but taken up in earnest in the early 1960s by Christopher Alexander (1964) and Jane Jacobs (1962) amongst others. This paradigm has taken a century in sensitising us to the need to step carefully when intervening in complex systems. Its message is that we plan 'at our peril' and that small interventions in a timely and opportune manner which are tuned to the local context are more likely to succeed than the massive top-down plans that were a feature of city planning throughout much of the twentieth century. To impress this new style of planning, we shall proceed by analogy using metaphors about how cities are formed taken from physics and biology. We shall first outline the notion of modularity and hierarchy, of self-similarity and scale in the physical and functional form of cities, and then we shall present ways in which basic functions generate patterns that fill space to different degrees. Cities develop by filling the space available to them in different ways, at different densities, and using different patterns to deliver the energy in terms of people and materials which enable their constituent parts to function. We shall demonstrate a simple diffusion model, and then generalise it to grow city forms and structures in silico. We shall allude to city plans in history that demonstrate our need to plan with and alongside the mechanisms of organic growth rather than against these processes, which has been the dominant style of planning in the past century.

3 Modularity, hierarchy and self-similarity

There is wonderful story first told by Herbert Simon (1962) which illustrates the importance of hierarchy and modularity in the construction of stable and sustainable systems. Simon tells of two Swiss watchmakers, Hora and Tempus, who both produced excellent but identical watches, each of a thousand parts. The key difference between the watchmakers was in the processes they used to produce each watch. Tempus, for example, built each watch by simply taking one part and adding it in the requisite order to the next until the whole assembly was complete. Hora, however, built up his watches in subassemblies, first of ten parts each. Once he had produced ten subassemblies, he added these into a large subsystem containing a hundred parts. When he had added all his ten-part assemblies into the larger parts of one hundred, he completed the whole watch by simply adding the ten larger assemblies together. It took Hora only a fraction longer to add the subassemblies. To all intents and purposes, the completed watches took the same time and were no different.

As the fame of the watchmakers grew, they received more and more orders; but, in this fictional world, the only way they could receive these orders was by telephone. Every time the telephone rang Tempus had to put down the watch, and it fell to pieces, so he had to start again once he had taken an order. Hora, on the other hand, simply lost the subassembly he was working on when the telephone rang and he had to put down the watch to answer it. You can see immediately what happened in the long term. Tempus found it more and more difficult to complete a watch as the telephone rang more and more, while Hora simply traded off his telephone time for watchmaking, for his process was robust. Ultimately Tempus went out of business while Hora prospered, and the moral of the story is that sustainable systems which can withstand continued interruptions of this kind are built in parts, from the bottom up, as modules to be assembled into a hierarchy of parts.

Modular construction is not simply a functional process of ensuring that component parts of a system are stuck together efficiently and sustainably but also a means of actually operating processes that drive the system in an effective way. For example, different functions which relate to how the economy of a city works depend on a critical mass of population such that the more specialised the function, the wider the population required to sustain it. In short, more specialised functions depend on economies of scale such that the size and spacing of various functions produces a regular patterning at different hierarchical levels. The modules are thus replicated in a way that they change their extent with their scale. We can demonstrate this point using some simple geometry that illustrates how we can scale a physical module, producing a fractal that is similar on all scales. Imagine that we need to increase the space required for planting a barrier along a straight path. If we divide the line of the path into three equal segments, we can take two of these segments and splay them away from the path so that they touch and form an equilateral triangle in the manner that we show in Figure 1.1(a). This clearly increases the length of the line L (which has an original length of three units) by one unit, so that the new length of the line becomes $(4/3)L$. We can further increase the length of the line by subdividing each segment into three and displacing the central portion of each of the original segments to form the same equilateral triangle but at a scale down from the original. If this is done for each of the original four segments, then the length of the second line composed of these four segments increases by $4/3$. This in turn is $4/3$ the length of the original line L, and the new line is now $(4/3)(4/3)L$. We can continue doing this at ever finer scales, and the length of the line at scale n thus becomes $(4/3)^n L$.

This construction is a recursion of the same rule at different scales, and it generates a pattern which is self-similar in that the motif – the triangular displacement – occurs at every scale and is in a sense the hallmark of the entire construction. The structure grown from the bottom up produces a shape that is a fractal, a regular geometry composed of irregular parts which are repeated on successive scales, which is indicative of the same processes being applied over and over again. The process can be viewed as a hierarchy which is clearly present in the pattern itself but, in terms of the recursive process, can be abstracted into the usual tree-like diagram which we show in Figure 1.2.

There are several strange consequences to the process that we have just illustrated. If the process of adding more and more detail of the same kind continues indefinitely, the length of the line increases to infinity, but it is intuitively obvious that the area enclosed by the resulting shape either in the Koch curve in Figure 1.1(a) or the Koch island in Figure 1.1(b) converges to a fixed value. Second, if the line becomes more and more convoluted in filling the plane, then it would appear that the line which has Euclidean dimension of 1 seems to have the dimension of the plane which is 2. This concept of space-filling can be formally demonstrated to be encapsulated in the idea of fractal dimension. The Koch curve in Figure 1.1(a) has a fractal dimension about

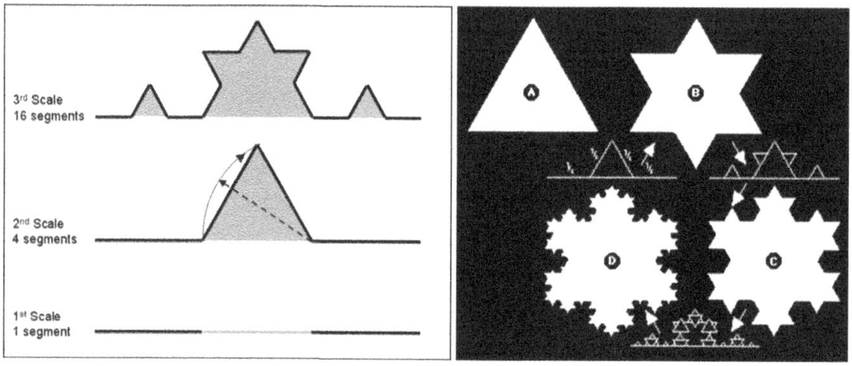

(a) (b)

Figure 1.1 Constructing a space-filling curve: the Koch snowflake curve: (*a*) successive displacement of the central section of a line at ever finer scales; (*b*) application of the displacement rule to the lines defining a triangle shape called a Koch Island

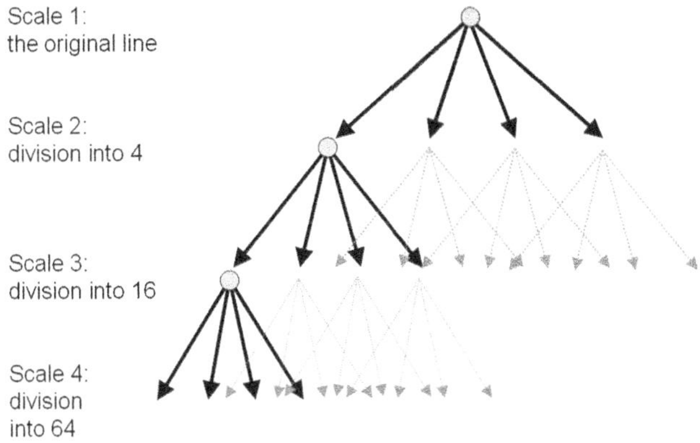

Figure 1.2 The hierarchy of composition in constructing a fractal

1.26 while a more convoluted line like a fjord coastline has something like 1.7. Rather smooth curves such as the coastline of southern Australia have a fractal dimension of about 1.1. In fact the inventor of the concept of fractals, Benoit Mandelbrot, wrote a famous paper in *Science* in 1967 which was entitled 'How long is the coast of Britain? Statistical self-similarity and fractional dimension'. To cut a very long story short, objects which are irregular in the way we have shown, and which manifest self-similarity, are fractals whose dimension lies between the dimension that they are defined by and the dimension of the space that they are trying to fill. In cities, filling the two-dimensional plane with particular forms of development from the parcel to the street line, and at different densities, suggests that their fractal dimension lies between 1 and 2. Thus this dimension becomes the signature of urban morphology, which is the outcome of processes that generate fractal shapes (Batty and Longley 1994).

There is, however, a much more literal morphology which is fractal, and this is the shape of an object or set of linked objects that form a tree or dendrite. If you want to transport energy from some central source to many distant locations, it is more efficient to develop infrastructure that captures as much capacity for transfer as near to the source as is possible. This is rather easy to demonstrate graphically, for if there are sixteen points arranged around a circle, then rather than build a link between the source and each of these sixteen points it is more efficient to group the links in such a way that the distance to these different locations is minimised. In Figure 1.3(a), assuming that each single link is of distance 1 unit, the length of the routes needed in total to service these locations ('fill the space') is sixteen in comparison with the grouping of these routes into two, then four, then eight, which is shown in Figure 1.3(b). The total distance of this arrangement in Figure 1.3(b) is something between a half and three-quarters of the original form in Figure 1.3(a) depending on the precise configuration, although the capacity of the links which take more traffic nearer the source are bigger, and this would incur extra costs of construction. Nevertheless, this demonstrates the important point that, where resources are to be conserved (which is in virtually every situation one might imagine), space must be filled efficiently. The tree structures

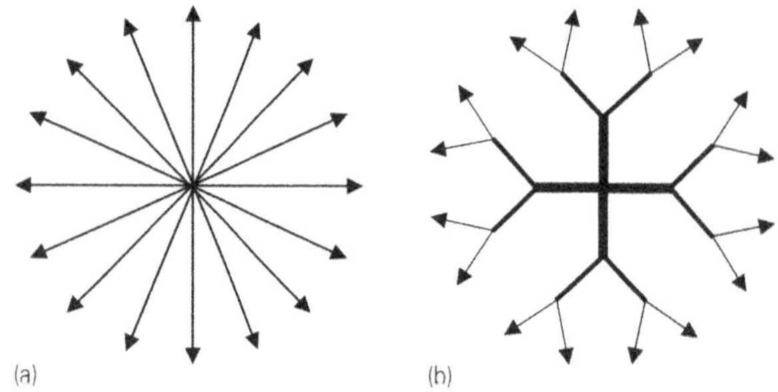

Figure 1.3 Literal hierarchies: transport from a central source. (*a*) Each link is separate; (*b*) arranging links into a more efficient structure

in Figure 1.3 are fractals, with Figure 1.3(b) illustrating this self-similarity directly while at the same time being a literal hierarchy spread out in space demonstrating quite explicitly the pattern of its construction.

There are many examples of such hierarchical structure in the forms we see in both nature and made-made systems. Energy in the form of blood, oxygen, and electric signals are delivered to the body through dendritic networks of arteries and veins, lungs, and nerves, as we illustrate in the schematic of the central lung system in Figure 1.4(a). Plants reach up to receive oxygen from the air and down to draw out other nutrients from the soil as in Figure 1.4(b). Nearer to our concern here, and reflecting the discussion of route systems above, Figure 1.4(c) shows the network of streets in the mid-size English town of Wolverhampton (population *circa* 300,000 in 2001). It is clear that the traditional street system has grown organically, but the ring around the town centre has been planned, imposed from the top down, thus illustrating the notion that what we observe in cities is a mixture of different scales of decision-making. In Figure 1.4(d), we show one of the Palm Islands off the coast of Dubai developed by the construction company Nakheel. This is a wonderful example of how it is necessary to conserve resources when building into hostile media – in this case by reclaiming land from the sea, where transportation and access become the main themes in the way the resort is formed.

These examples demonstrate that cities are built from the bottom up, incrementally, and where they are planned from the top down the plan is usually a small part of wider organic development. When cities grow at any point in time, we have little idea of what the future holds with respect to new behaviours, values, technologies and social norms, and thus it is not surprising that cities grow in an *ad hoc* manner reflecting the efficiencies and equities that dominate the consensus at the time when development takes place. To illustrate how we can model this process, we can abstract it into two main forces that reflect the desire for space on the part of any individual, developer or consumer which is traded off against the desire to live as close as possible to the 'city' composed of other individuals, so that economies of scale might be realised. This is a simple model which captures all the ideas we have introduced so far, and we shall now develop it as a hypothetical simulation.

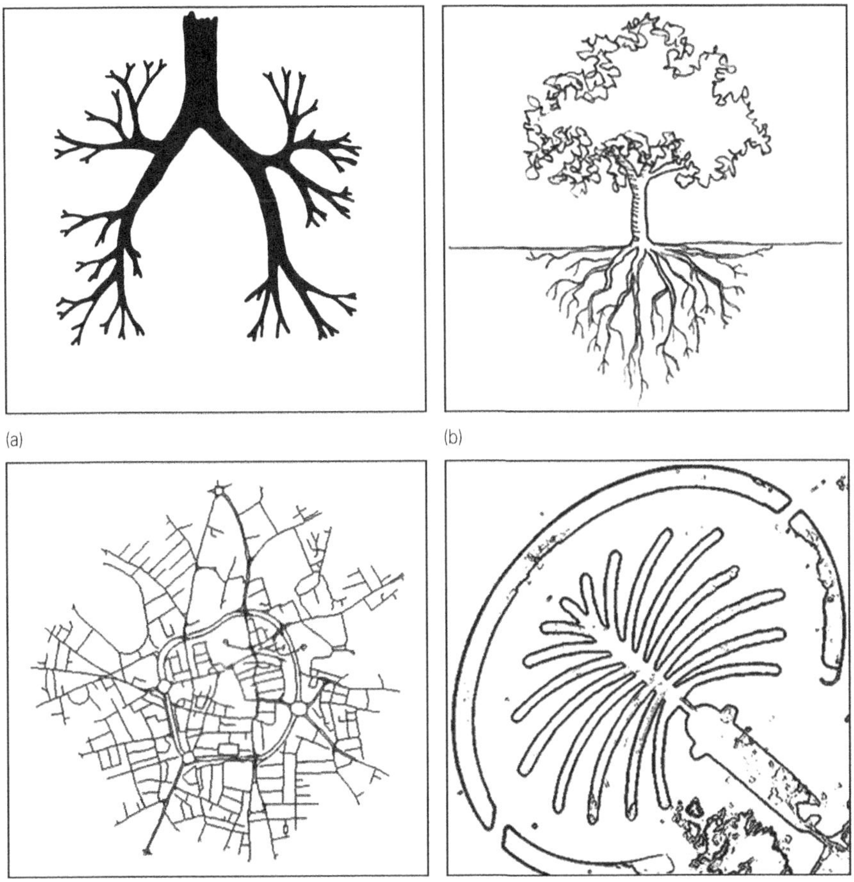

(a) (b) (c) (d)

Figure 1.4 Space-filling hierarchies: (*a*) a schematic of the human lung; (*b*) a schematic of a tree growing into different media – air above ground and soil below; (*c*) the road network of a mid-sized English town, Wolverhampton; (*d*) space-filling in difficult media – Nakheel's Palm Island in Dubai

4 Simulating space-filling growth

Our model is based on two key drivers. First, we would all agree that cities exist as machines for enabling us to divide our labour so that we might realise economies of scale – or agglomeration economies as they are increasingly called. Alfred Marshall made the point over a hundred years ago:

> Great are the advantages which people following the same skilled trade get from near neighbourhood to one another. The mysteries of the trade become no mystery but are, as it were, in the air.
>
> (quoted in Glaeser 1996)

Our first principle is that individuals must be connected to one another in terms of their proximity to others for the city to exist, and this means that new entrants to the city must somehow connect physically to those already there. In contrast, individuals seek as much personal space as possible for themselves, and this translates into the notion that they wish to live as a far away from others as possible in the city space. This may translate into living at low densities; but, as in Manhattan, large apartments in the sky may be another way of realising this quest, while there are increasingly innovative ways of meeting this goal by specialising in living in different locations. In our context here, we shall embody this second principle as one where people wish to live on the edge of the existing city rather than in the centre, notwithstanding the great variety in these kinds of preference.

Our model can be constructed as follows. Imagine that a trader decides to locate his or her base at the intersection of a trading route and a river where the land is fertile and flat. Many cities have grown from such humble origins where comparative natural advantages such as these determine the best location for settlement. Now imagine that another individual seeking a permanent location wanders into the vicinity of the lone trader's base. If that trader happens by chance to get within the neighbourhood of the existing trader, that trader decides to locate there, although there may be many traders in the wider hinterland who do not enter into the neighbourhood and never find the emergent settlement. However, a certain proportion will find the settlement; and with a certain probability, and given enough time and traders, the settlement will grow. From these simple principles, we can demonstrate the form of the growing city. In Figure 1.5(a), we show a schematic of the location process. The individuals are arranged around a circle well outside the location of the settlement, which is fixed at the centre of the circle with the dark solid dot. This is where the original trader locates. Each individual is a solid light dot, and they begin the movement in search of the location using a random walk. They decide at each step to move up or down or left or right randomly, and in this way walk across the locational plane. If they move to a cell adjacent to the fixed solid dark dot, they settle; they stop any further walking and turn dark which shows that they are now fixed, stable and no longer in motion. The first one to do so is shown by the shaded dark dot adjacent to the initial dark dot. That is all there is to it. You can see the final form, so you know what will result; but if you had not seen this result, then many would guess that the result would not be a tree-like structure but a compact growing mass.

We show the progression of this in Figure 1.5(b); and, to talk this through, what happens is that, as soon as a trader settles next door to the existing dark dot, the chances of another trader finding that new trader settlement as opposed to any other one increase just a fraction. As time goes by, the linear edge pattern that is characteristic of the growing tip of the cluster begins to emphasise itself, and increasingly a trader finds it impossible to penetrate the fissures in the growing cluster. Traders then are more likely to find the growing cluster at its edge, and in this way the cluster begins to span the space. If this were to produce a growing compact mass, then it would have dimension nearer to 2 – the Euclidean dimension – but in fact it has a fractal dimension between 1 and 2, about 1.7, so empirical work in many fields has determined. This, like any dendritic structure, is a fractal, and it is easy to see the self-similarity that is contained in its form. Break off any branch and you can see the entire structure in the branch just as you can usually see the entire structure of a tree or a plant in its leaf. As we increase the resolution of the grid or lattice on which this walk

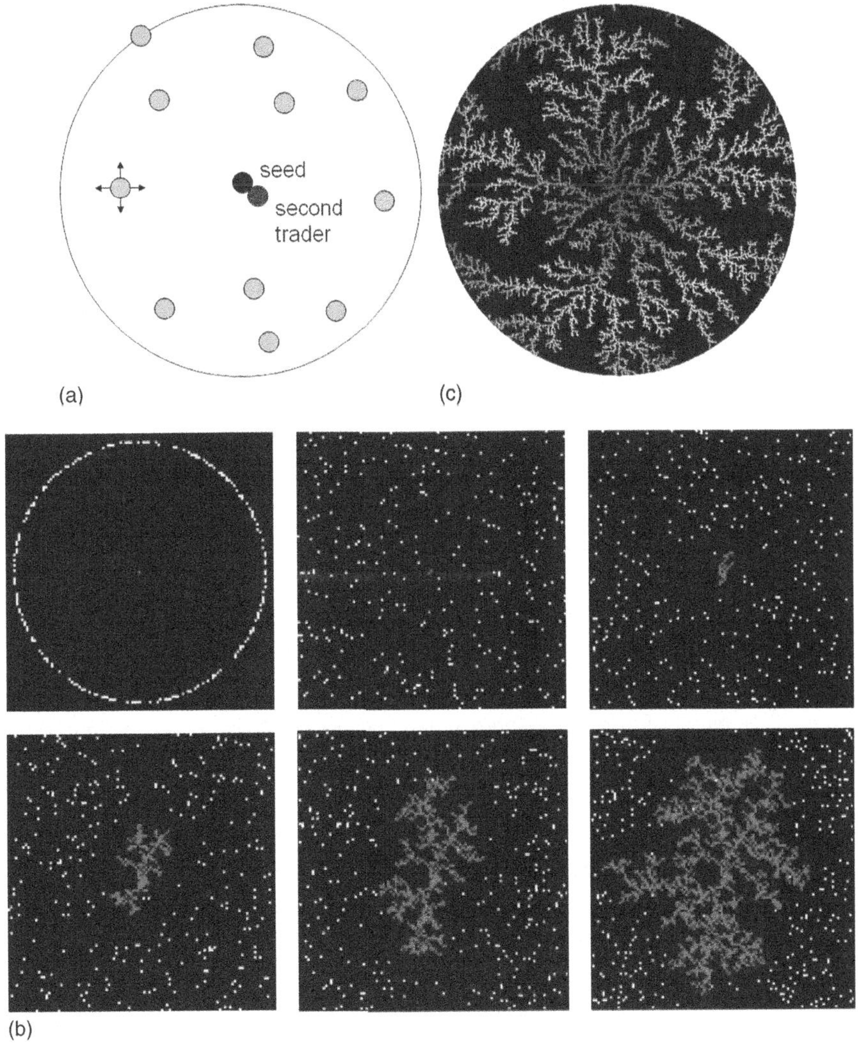

Figure 1.5 Generating clustered city growth using diffusion-limited aggregation (a) The dark dot represents the fixed location of the first actor forming the city while the light dots represent actors searching for sufficient agglomeration economies to become fixed to the city (b) the process of becoming fixed to the growing city which is the dark mass (c) the ultimate city form which is generated in dendritic form

takes place, we get finer and finer tree-like structures where the fractal structure is readily apparent, as we show in Figure 1.5(c).

This form is generated by a process called diffusion-limited aggregation (DLA), which has been found and used extensively in physics to grow crystal-like structures and to examine ways in which one medium penetrates another, such as oil diffusing into water. In fact you can see similar patterns if you pour concentrated liquid soap into ordinary bathwater, and this is also reminiscent of the way the Dubai Palm Islands resort has been 'forced' into the sea. It is a general principle that a substance with a higher density creates such patterns when infused into a substance of a lower density. A model, of course, is only as good as its assumptions, but it is possible to tune this DLA model to produce many different shapes, some of which bear an uncanny resemblance to those that we find in real cities. For example, we can 'tune' the DLA to produce sparser structures if we relax the criterion that the individual who settles must exactly touch the already settled structure. We could, for example, set a distance threshold for this, or we could insist that more than one individual must already have settled. In this way, we can change the density, growing structures that are very heavily controlled in their dependence on what has gone before, generating linear structures all the way to compact ones where the degree of control over where traders are allowed to settle is very weak. Some of these extensions are illustrated in my book *Cities and Complexity* (Batty 2005).

5 Real world cities and patterns of complexity

There are many examples at different scales of the way cities are structured along dendritic lines mirroring the lines of energy that serve their distant parts. We glimpsed an idea of this in Figure 1.4(c) where we abstracted the street network of Wolverhampton, but cities are not pure dendrites. Different networks are superimposed on one another for different kinds of transport ranging from different modes requiring different networks through to social and electronic networks which underpin the way people trade and communicate. In Figure 1.6, we show a map of inner London where the streets are grey-scaled according to the energy they transport, in fact using the proxy of road traffic volumes which give some index of both capacity and congestion or saturation. This is also highly correlated with patterns of accessibility which mirror the proximity of places to each other.

Street networks are excellent examples of how cities grow from the bottom up, for they represent the skeletal structure on which all else in the city hangs. As we can see from the way cities grow, transport and land use are intimately related. Indeed, in the 1930s, as urban sprawl first became significant in Britain, 'ribbon' development became the pattern of transport linked to land use that was the subject of fierce control in the quest to contain urban growth. It is possible to see this connectivity in many patterns of urban growth. The picture of the way the eastern US city of Baltimore in Maryland has grown over the last 200 years that we show in Figure 1.7 is a clear illustration of the way development proceeds along radial routes from the traditional centre, particularly at the turn of the last century when streetcars, bus and railways systems dominated. Although this pattern is breaking down as cities become more polycentric and specialised in their parts, and as new kinds of central business district such as 'edge cities' become established, it is still significant.

These patterns recur at different scales, although the notion of them being faithfully reproduced at every spatial scale needs to be tempered with the obvious fact that individuals are diverse in their tastes and values, and thus heterogeneous in their actions. Moreover the sort of similarity that occurs in cities is statistical self-similarity, not the rather strict self-similarity that we saw for example in the construction of the Koch snowflake curve in Figure 1.1. In fact, although the pattern of transport routes in cities is generally radial, focusing on significant hubs, and organised according to a

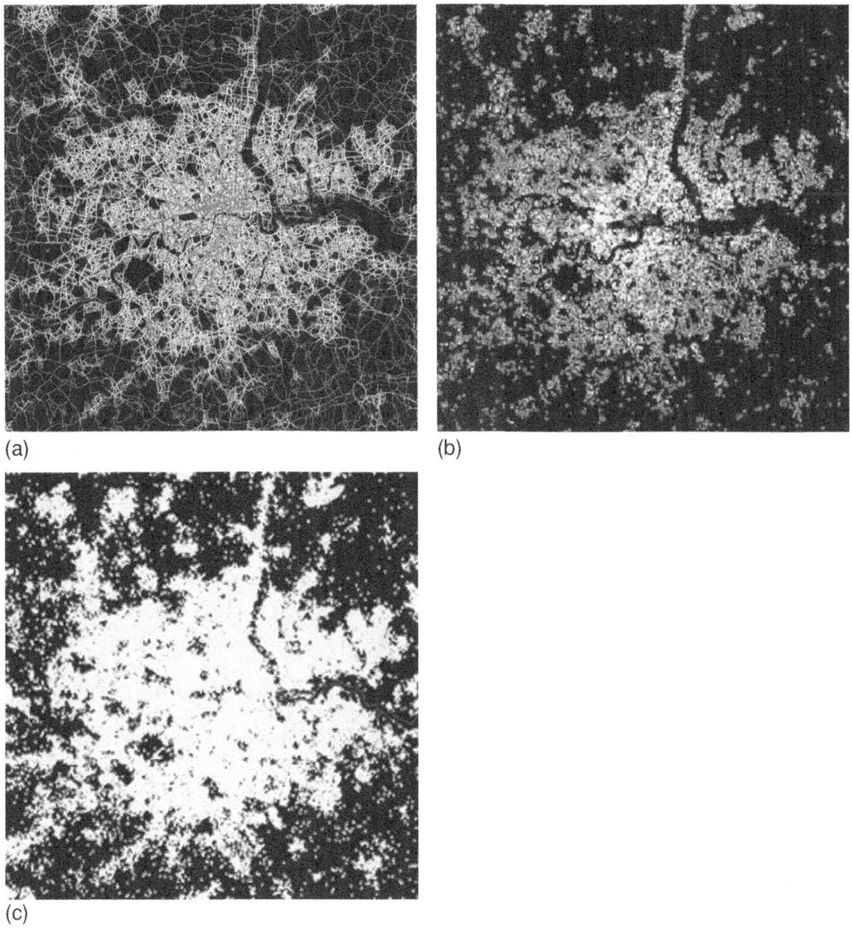

Figure 1.6 The organically evolving network of surface streets in Greater London classified by traffic volume (a) The street network in Greater London shaded according to distance from the centre in 2008 (b) population density in Greater London in 1991 (c) night lights defining Greater London from space in 2008

hierarchy of importance which mirrors different transport technologies, capacities and speed of transmission, street systems illustrate the space-filling principle quite clearly. At the local level, there is more conscious planning and design of street systems, particularly in developments which are self-contained, for purposes of the actual construction themselves as well as their financing and sales. For example, residential areas are often formed as small single-entry streets into houses arranged around cul-de-sacs for purposes of containment and traffic management as well as security. We shall return to these ideas in the next section when we speculate on how such structures can be formed in a more conscious sense through explicit design and planning; but it is important to note that these patterns do recur across different scales, as can be seen in their statistical distributions as well as in their physical self-similarity.

The obvious question is how far can we get with the diffusion-limited aggregation model of the last section in generating simulations of real structures that we see in the transport development of London in Figure 1.6 and urban development of Baltimore in Figure 1.7? This kind of model is of course a demonstration of how two

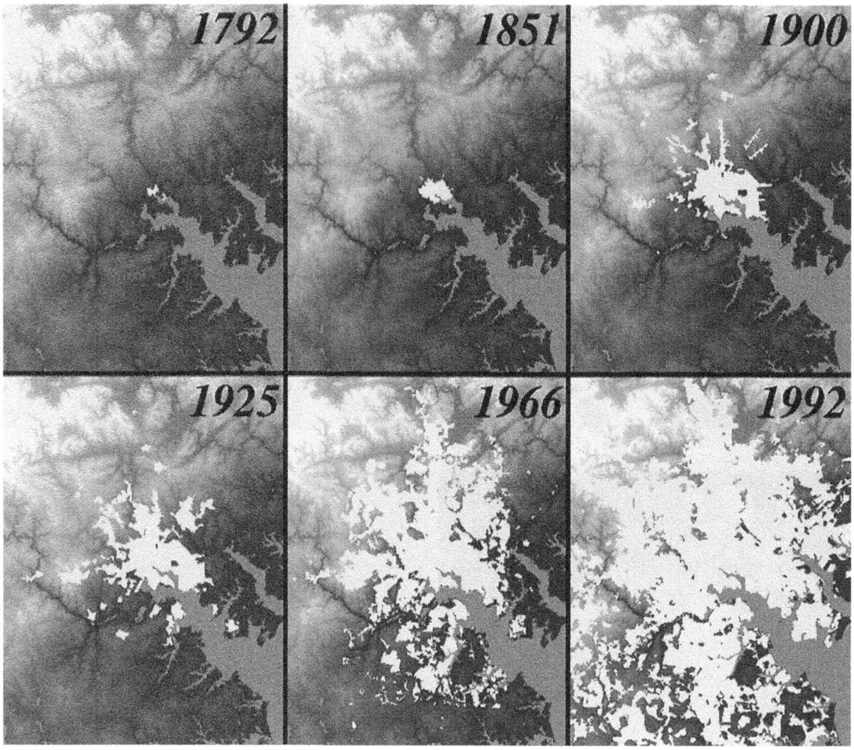

Figure 1.7 The two hundred years of urban growth in Baltimore. Source: http://landcover.usgs.gov/LCI/urban/umap/pubs/asprs_wma.php

principles or forces interact to produce a structure that resembles certain features of the modern city. It is not intended as anything other than a graphic way of impressing the notion that bottom-up uncoordinated change leads to highly ordered structures – fractals – which emerge from this comparatively simple process. One can begin to illustrate how one might make this more realistic, but it is a far cry from the kinds of operational models that are used routinely for strategic planning by government and other agencies. The model is made more realistic simply by planting it into a space or terrain that has real features. In Figure 1.8, we show four different simulations of development in the town of Cardiff in Wales which takes the coastline and rivers which define that area. We set two seeds, one at the historic centre and one at the dockside, and let the DLA model operate in the manner we have shown in Figure 1.5. From this, we realise quite quickly that the river cutting the town in two makes a difference to the rate of growth in parts of the town, while the fact that Cardiff has two centres shows how difficult it is to generate a pattern that gives the right historical balance to each. This is not surprising because none of the factors that affected this competition between the two centres is contained in the model. A more detailed discussion of the simulation is presented in our book *Fractal Cities* (Batty and Longley 1994). The model, however, can only simulate patterns that are a consequence of its assumptions. Yet these kinds of simulation also provide a means of demonstrating and testing various future hypotheses about urban form. By way of providing some sense of closure to

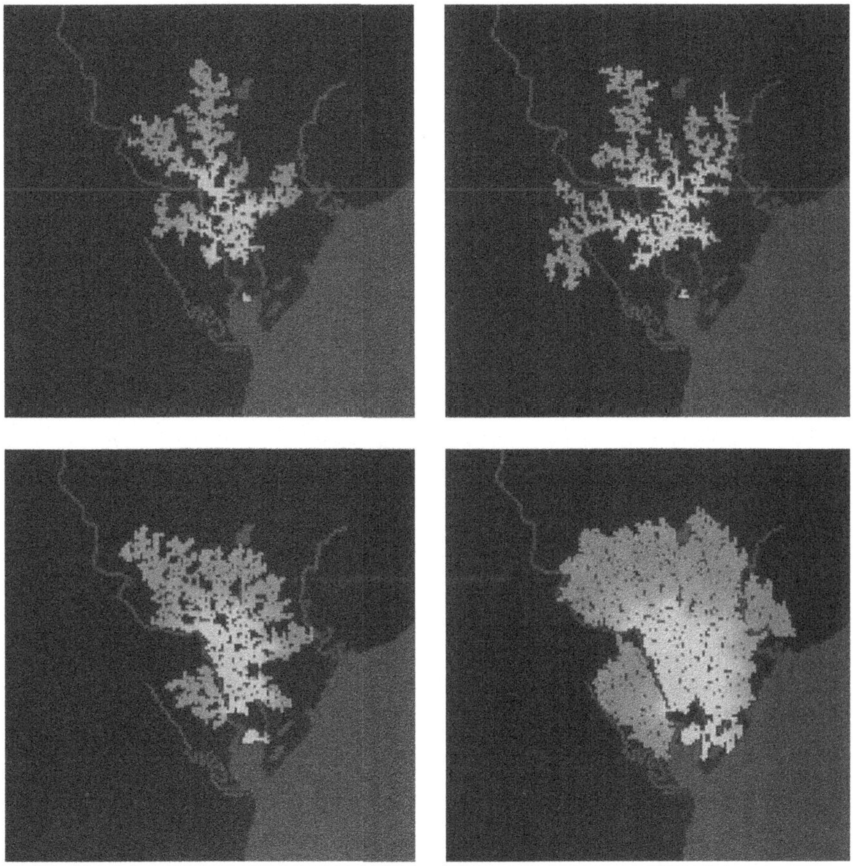

Figure 1.8 Simulating growth using DLA in the spatial landscape, centred on the city of Cardiff

this chapter, we shall show how such models can be used to help to generate effective designs.

The control over the development is tuned to decrease systematically through the simulations from top left to bottom left in clockwise order.

6 Generating idealised cities

We need a greater degree of control over our simulation process than that provided by the diffusion-limited aggregation model or its variants. In fact the way we generated the previous clusters is using a generative algebra that lies at the basis of many pattern-making procedures called automata. Automata are usually defined rather generally as finite-state machines driven by inputs which switch the states of the machine – the outputs – to different values. The outputs from the machine may then be used as inputs to drive the process of state transition through time, and this generative process can be tuned to replicate the sorts of pattern that we have been discussing in this essay. For example, the input to the DLA model is an individual who moves in a cell space; and, if certain conditions in the space occur, the individual changes the state of the cell from undeveloped to developed. This is, of course, done in parallel for many individuals. The idea that the space might be characterised as a set of cells simply gives some geometric structure to the problem; and, although we have taken for granted the fact that cities are represented in this way in these simulations, for automata in general, and spatial automata in particular, they can be any shape and in any dimension.

The automata we use here to generate physical development are called cellular automata (CA) where we assume a regular lattice of (square) cells in which development takes place by changing the state of each cell from undeveloped to developed as long as certain rules apply. The elements of CA are thus: a set of *cells* which can take on one of several states, in this case developed or undeveloped, extendable to different kinds of development; a *neighbourhood* of eight cells in the N-S-E-W-NE-SE-SW-NW positions around each cell in question; and a set of *transition rules* that define how any cell should change its state dependent upon the configuration, state and possible attributes of cells that exist within the neighbourhood of the cell in question. Now, if we apply this model starting with the initial condition of one cell in the centre of the lattice being switched on – developed – and apply the rule that, if there exists one or more cells in the neighbourhood of any cell, this will generate a diffusion around the initial cell which mirrors the process of successive spreading of the phenomena, just as a physical substance with some motion might diffuse. The diffusion is square because the underlying lattice is square, but we can easily develop versions in which the diffusion is circular if we so configure the lattice. We show this diffusion and the rules of engagement in Figure 1.9. If we then modify the rules by noting that the number of cells in the neighbourhood of an undeveloped cell must be only one, then we generate the diffusion in Figure 1.10(a); and if there are one or two, then the simulation generates Figure 1.10(b). There are literally millions of possibilities, and the trick is, of course, to define the correct or appropriate set of rules. Wolfram (2002), in his book *A New Kind of Science*, argues that such automata represent the fundamental units on which our universe is constructed. Although we have more modest ambitions

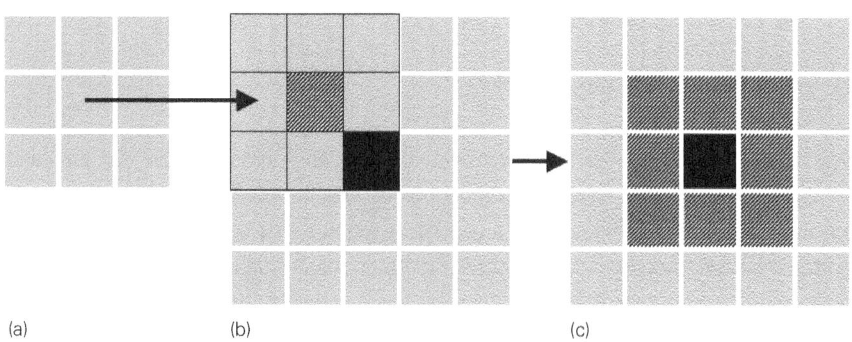

Figure 1.9 How cells are developed (*a*) An eight-cell neighbourhood around a central cell in question is applied to (*b*) each cell in a lattice. If one or more cells in the lattice is of a particular state, in this case developed (black), the state cell in question in the neighbourhood (black hatch) changes to developed. If this rule based on one or more cells in the neighbourhood is applied to every cell in the lattice, the result is (*c*) the set of cells around the central cell (black) become developed (black hatch)

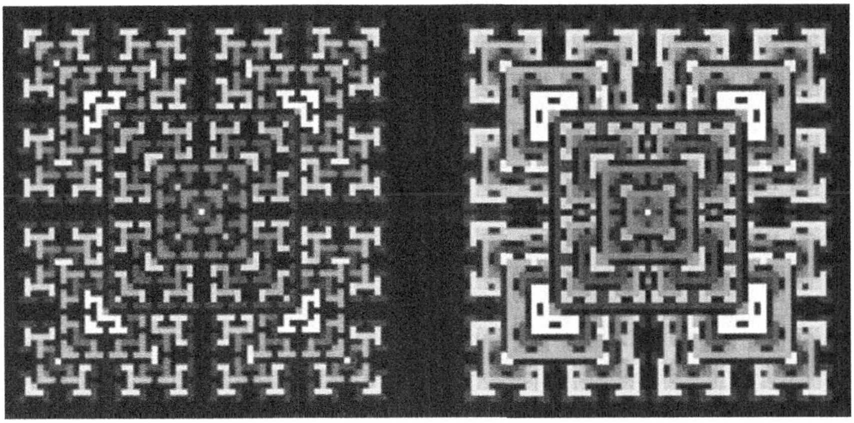

(a) (b)

Figure 1.10 Regular diffusion using CA: patterns reminiscent of idealised Renaissance city plans: (*a*) with only one cell in the neighbourhood; (*b*) with one or two cells developed

here, this kind of automata can be tuned to replicate many different generative phenomena which characterise many different forms of city.

To generate ideal cities using such automata, it is necessary to begin with a set of realistic rules for transition. Ideal cities are often designed to meet some overriding objective function – to minimise density as in Frank Lloyd Wright's BroadAcre city, to maximise density as in Le Corbusier's Cité Radieuse, to generate formal vistas and garden squares as in Regency London, to generate medium-density new towns with segregated land uses as in the first generation of British New Towns, and so on. A rather good example which can be generated using cellular automata principles is the plan for the Georgian colony of Savannah in the New World. Developed in 1733 by General James Oglethorpe, we show the plan in Figure 1.11; the CA rules might be imagined in analogy to the way we generate development in Figures 1.9 and 1.10.

Usually plans for ideal cities are not grown using a generative logic because plans are conceived of all-in-one-piece, so-to-speak, and the notion of an uncertain future is never in the frame. However, CA allows us to generate plans that evolve

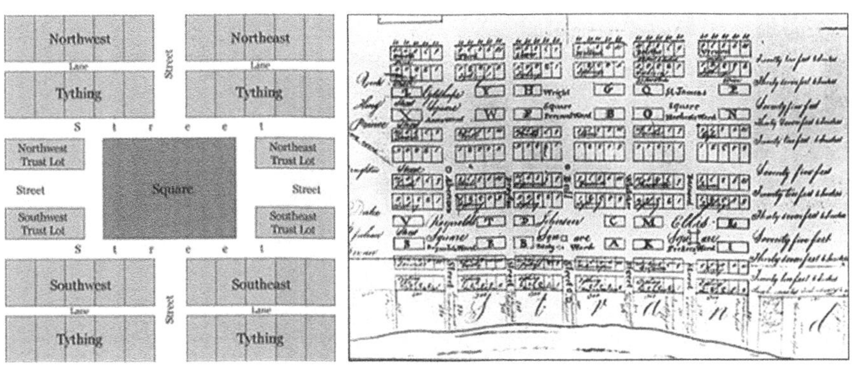

(a) (b)

Figure 1.11 The colonial plan for Savannah, Georgia: (*a*) the original neighbourhood plan; (*b*) the plan in 1770. Source: http://en.wikipedia.org/wiki/Squares_of_Savannah,_Georgia

through time, and we can continually change the rules, so that the idealisation is a shifting vision. In a sense, the plans which are grown in Figures 1.9 and 1.10 have stable rules which may or may not be considered as ideal objectives to be attained. To conclude our demonstration of this kind of logic and the intrinsic complexity of cities in that their ideal form is never certain, we shall return to the DLA model and tweak the rules a little so that a system-wide objective might be met. Imagine that the agents in our model move randomly in the manner we described earlier, that is to all points of the compass; this can be simulated using CA by assuming that where the state of the cell is an agent the cell changes state according to the movement. If an agent is at cell i, j, and it moves to cell $i + 1, j$ in the next time-period, then the cell state switches accordingly, from the cell where the agent is located to the cell where it is newly located. Our first rule, then, is simply cell state switching from the place where the agent was located to its new location. But we also have a rule that says that if the agent is located at cell i, j and there is another agent fixed at a cell in the neighbourhood of i, j, then the agent remains fixed and the cell on which it sites changes to the stable state. Note that in this version of the CA the cells contain mobile or fixed (stable) agents or have no agents within them at all. The cells have three possible states, which are appropriately coded, but this is still a CA with two sets of rules.

Now imagine that there is a strong wind blowing from south-west to north-east, and therefore any agent which is on the windward side of an occupied cell will not fix itself there. So, whenever an agent comes within contact of an agent already fixed on its leeward side, it continues to be mobile. Thus the development moves continually away from the point where the first agent locates, and what happens is that a line of cells is established across the space. In fact it is quite hard to guess what happens, and it is necessary to run the simulation to see what ultimate form the model might be. We show this in Figure 1.12, which is the picture of the city formed when the two principles of contact to the existing agglomeration and the need for as much space as possible are linked to the general objective of locating on the leeward side of already existing development. CA shows how this objective might be met.

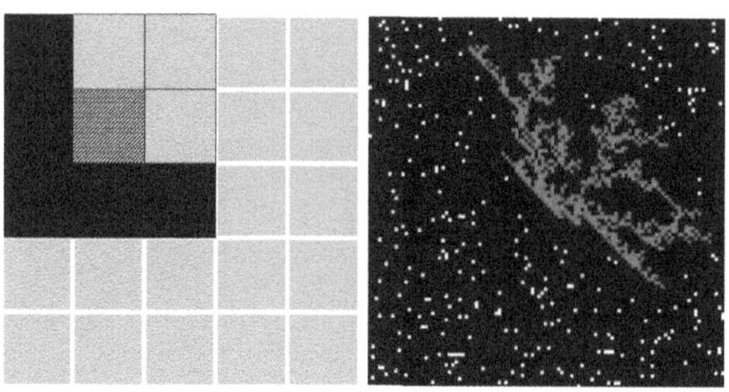

(a) (b)

Figure 1.12 Optimising growth by DLA: new development on leeward side of existing development: (*a*) if cells in black on the windward side are developed, the central neighbourhood cell is developed; (*b*) a typical outcome of the generative process

7 Next steps

There is much still to say about how cities are formed and evolve, how we might best understand and then simulate them and, most importantly, how we should design plans which enable them to function in more efficient and equitable ways. This essay has broached the idea that cities evolve into an unknowable future that is always uncertain. Therefore any goals that we might have for the future city are contingent on the present, hence continually subject to revision and compromise. In the past, cities have been designed in a timeless future where sets of objectives have been defined to be achievable as if the city were cast in a timeless web, and it is little surprise that few cities have ever achieved the aspirations set out in their plans. Complexity theory broaches the problem of the unknowable future and the way in which cities evolve from the bottom up, incrementally as the products of decisions that might be optimal at any one time but always subject to changing circumstances. This would appear to be a far more fruitful and realistic way of generating cities that meet certain goals, with the goals continually under review as the city emerges from the product of decisions which might be optimal in the small but whose global effects are unknowable in the large, until they emerge.

There are ways in which the processes that we have introduced here might be steered in more centralised ways, and it is the challenge of thinking in these terms that the complexity sciences are attempting to grasp: how control and management, planning and design, which traditionally have been configured and treated from the top down, might best be meshed with systems that grow and evolve from the bottom up. The answers probably lie in notions about hierarchy and the extent to which we might intervene and manage processes that generate hierarchies organically from the bottom up (Batty 2006). As we learn more about how cities evolve in these ways, it is my contention that we shall learn to plan less as we identify points of pressure and leverage. There, effective intervention and design in small, incremental ways might lead to large and effective changes that go with the flow, and do not fight against the grain. Such planning through incremental evolution has not been the history of most city plans hitherto, but our science is evolving to meet this challenge.

References

Alexander, C. (1964), *Notes on the Synthesis of Form*, Cambridge, Mass.: Harvard University Press.

Batty, M. (2005), *Cities and Complexity: Understanding Cities with Cellular Automata, Agent-Based Models, and Fractals*, Cambridge, Mass.: MIT Press.

Batty, M. (2006), 'Hierarchy in cities and city systems', in D. Pumain (ed.), *Hierarchy in the Natural and Social Sciences*, pp. 143–68, Dordrecht: Springer.

Batty, M. and Longley, P. A. (1994), *Fractal Cities: A Geometry of Form and Function*, San Diego, Calif.: Academic Press.

Geddes, P. (1915), *Cities in Evolution*, London: Williams & Norgate.

Glaeser, E. L. (1996), 'Why economists still like cities', *City Journal*, 6 (2), online at http://www.city-journal.org/

Jacobs, J. (1962), *The Death and Life of Great American Cities*, New York: Random House.

Mandelbrot, B. B. (1967), 'How long is the coast of Britain? Statistical self-similarity and fractional dimension', *Science*, 156 (3775): 636–8.

Simon, H. A. (1962), 'The architecture of complexity', *Proceedings of the American Philosophical Society*, 106: 467–82.

Wolfram, S. (2002), *A New Kind of Science*, Champaign, Ill.: Wolfram Media.

2 Embracing complexity in building design

Leonard R. Bachman

1 Introduction

We live in an emerging era of postindustrial society characterised by knowledge-based production of value, scenario planning of long-term future outcomes, multiple-stakeholder participation in projects, and intensely complicated design challenges. For these and related reasons, complexity in its deepest and most dynamic nature has progressively become a focal driver of building design. Consequently, as this chapter elaborates, the challenges of our design disciplines, theories, tools, discourse and responsibilities are progressively more critical, technical and vital. Given this broad perspective, the implications of complexity are not restricted to building science, but encompass both formal and performal criteria, technē and art, pragmatics and poetry. Complexity is infused in our societal institutions and cultural values; and it will certainly continue to impact our formulation of buildings. The point here is that, with this inevitable complexity as a given, our attention must turn to how design can positively accommodate the reality of systemic complexity to the significant human experience of the built environment.

2 History

First, to anchor the influences of complexity in an applied and relevant way, some postindustrial evolutions of professions in general (Table 2.1) and of design professions specifically (Table 2.2) illustrate the breadth and depth of impact. To risk a tautology here, the change is systemic. In every dimension of every pursuit, our emerging awareness of *reality as a dynamic system of macro- and micro-scale interactions* is replacing the narrower industrial age mentality – that of *manipulations that have only local and*

Table 2.1 Postindustrial professions (after Bachman 2003)

	Evolution of postindustrial professions
PHYSICS	Quantum mechanics and the Unified Field Theory (Bohr, Heisenberg, Hawking)
ENGINEERING	Non-linear and chaotic systems
PSYCHOLOGY	Self-actualisation and psychosynthesis (Maslow, Graf)
SOCIOLOGY	Knowledge-based culture (Kuhn, Bell)
BUSINESS	Industrial Organisational Psychology
MEDICINE	Holistic health and mind/body healing (Chopra)
AGRICULTURE	Organic gardening and beneficial insects (Rodale)
ECONOMICS	Life-cycle costs and externalised accounting (Henderson)

Table 2.2 The overlay of industrial and postindustrial society in the context of building design (after Bachman 2006)

	INDUSTRIAL ESTABLISHMENT	POSTINDUSTRIAL EMERGENCE
PLANNING	Survival sustenance from nature	Ecological sustainability with nature
	Anthropocentric	Biocentric
	Linear production	Cyclical flows
	Tactical objectives	Strategic goals
	Short-term plan	Long-term scenario
	Incremental shifts	Continuous change
PRACTICE	Product- and tradition-oriented	Process- and discipline-oriented
	Local effects of action	Global effects of interaction
	Mechanistic relationships	Systemic relationships
	Machine as the icon	Nature as the icon
	Heuristic procedures	Cybernetic integration
	Physical prototype modelling	Analog simulation modelling
	Mass standardisation	Mass customisation
DESIGN	Hierarchical and linear	Holistic and non-linear
	Embrace deterministic simplicity	Embrace teleological complexity
	Intuitive heuristics of form	Self-emergent intelligent form
	Anticipate the inevitable future	Design of future scenarios
	Innovative individuals	Transdisciplinary teams
	Pioneer-as-hero model	Designer-as-collaborator model
	Design for élite status	Design for social justice
	Manual and automatic control	Intelligent automation
	Transient static solutions	Robust dynamic solutions

immediate consequence. Complexity in building design is clearly a natural extension of this new thinking.

Lyle (1994: 13) argued for the many writings and collaborations of Patrick Geddes and Lewis Mumford as a genesis in this evolution in design (e.g. Geddes 1904; Mumford 1922). Lyle identifies their joint designation of 'eotechnic, paleotechnic, and neotechnic' eras in the relation of technology to society as the respective equivalents of prehistory, industrial times, and the presages of postindustrial society. For Lyle the distinction is particularly useful in understanding the transition away from the simplistic paleotechnic linear throughput model beginning with raw resources, to manufactured consumer goods, to used-up waste pollution. In its stead Geddes and Mumford are credited with foreseeing the neotechnic advent of ecologically oriented design as participation in natural cycles of growth, decay and regeneration. As a precursor of complex buildings, the same neotechnic principles can be applied beyond sustainable regenerative design to any factor of dynamic interaction.

To situate this building design evolution in more current history, consider how the following precepts and trends are manifested in today's building design (Bachman 2003):

- Bauhaus principles of wedding artisan quality design with industrial mass production
- Brutalism and the advent of exposed and expressed building systems
- Trends from handmade buildings to prefabricated kit-of-parts assembly
- Ascendance of critical technical issues in concert with formal issues
- Emergence of environmental issues over structural ones
- Reliance on optimisation tools over intuitive accepted practice
- Transition from monumental affect to sustainable effect

Each of these transitions has moved building design into progressively more complex methods and tools, and each has resulted in a more holistic and complex definition of what a building is in the first place. As a subtext, each of these trends has contributed to the understanding of building design as an exercise in systems thinking at many levels. First, there are the building systems themselves, each acting as an ordered set of flows: structure as the flow of gravity, stress and strain for example; or circulation as the flow of people, ventilation as the flow of air, daylighting as the flow of natural illumination, and so forth. Next there is the concept of the building design as a system of solutions: multiple stakeholders with different and conflicting expectations, multiple disciplines with differing expertise, multiple and overlapping accountabilities for the measure of success, and so on. There is also the idea of the design itself as a resolution of intentions, resources, constraints and opportunities. Beyond that, the systems basis of building design is also reinforced by issues of sustainability, robust flexibility and adaptation, multimodal response to seasonal and daily climate conditions, and a host of other challenges that cannot be satisfied with intuitive levels of fitness or static formalistic propositions. A further account of the relation of systems and complexity is covered in the next section of this chapter.

A more episodic and storied recounting of these trends seems to have its epicentre in the 1960s. It is no accident that so many of the events that triggered an awareness of the true complexity of buildings arose then, because that decade was part of the transition away from a growing dissatisfaction with industrial society and a shift to the postindustrial consciousness of global, environmental and macro-scale interrelatedness. Table 2.3 is an abbreviated list of the more significant publications of the 1960s that are associated with these events.

Several items on this 1960s hit-list are still important to building designers, including Jacobs's *The Death and Life of Great American Cities* (1961), Alexander's *Notes*

Table 2.3 Major publications of the 1960s recognising complexity in building design

Complex buildings: some pivotal publications of the 1960s		
DATE	**AUTHOR**	**TITLE**
1961	Jacobs	The Death and Life of Great American Cities
1962	Carson	Silent Spring
1963	Ackoff	A Manager's Guide to Operations Research
1963	Fuller	Operating Manual for Spaceship Earth
1963	Olgyay	Design with Climate: Bioclimatic Approach to Architectural Regionalism
1964	Alexander	Notes on the Synthesis of Form
1965	Kepes	Structure in Art and in Science
1965	Simon	The Shape of Automation for Men and Management
1966	Alexander	The City As a Mechanism for Sustaining Human Contact
1966	Kepes	The Man-Made Object
1966	Norberg-Schulz	Intentions in Architecture
1966	Venturi	Complexity and Contradiction in Architecture
1967	Mumford	The Myth of the Machine
1968	Ackoff	Fundamentals of Operations Research
1968	Bertalanffy	General System Theory: Foundations, Development, Applications
1968	Churchman	The Systems Approach
1968	Sanoff	Techniques of Evaluation for Designers
1969	Arnheim	Visual Thinking
1969	Fuller	Utopia or Oblivion: The Prospects for Humanity
1969	Givoni	Man, Climate, and Architecture
1969	Jacobs	The Economy of Cities
1969	Kahn	The Year 2000: A Framework for Speculation on the Next Thirty-Three Years
1969	Knowles	Owens Valley Study: A Natural Ecological Framework for Settlement
1969	McHarg	Design with Nature
1969	Peña	Problem Seeking: New Directions in Architectural Programming
1969	Sanoff	A Systematic Approach to Design
1969	Simon	The Sciences of the Artificial

on the Synthesis of Form (1964), and others. Of particular note to this point of the discussion, however, is William Peña's *Problem Seeking* (1969) as a demarcation in the early transition from intuitive design expertise and genius talent models to the knowledge-based and collaborative team model of postindustrial complexity. While there are many other entries in this field (e.g. Sanoff, Preiser, Hershberger), *Problem Seeking* set new standards for how designers might construct a problem space and identify the essences of what they were designing as well as who they were designing for. It remains the best-known primer on knowledge-based design in architectural programming, and Peña's disciples are still at the fore of this field today (e.g. Cherry 1999).

3 Systems theory

Some discussion is offered here to connect the notion of complexity to that of buildings and building design. A complex dynamic system, on the one hand, is understood to have certain inherent characteristics which spontaneously result in purposeful behaviour such as adaptive feedback, self-organisation or robust response to disturbance. What always drives the action of such systems is potential, meaning that there is some set of opposing or dialectic forces to which the system responds in these behavioural ways.

Buildings, on the other hand, may at first seem like static objects with a few moving machine parts. On further consideration, however, we realise that they, too, respond in complex ways to several sets of bimodal states which might be considered dialectic in how they compel adaptive responses. There are several such temporally situated pairs of states, such as summer versus winter weather, day versus night, moving traces of the sun, aerodynamic response to winds and breezes from different vectors. Other potential energies include occupied and non-occupied states, dead and live structural loads, flexibility of use, adaptability of function, and so forth.

Buildings which embrace these bimodal states do not simply suffer the changing conditions in benign response. On the contrary: responsive, adaptive, systems-based design admits winter sun and shades other periods when cool air is desired; it allows pleasant breezes and redirects winter winds with bimodal aerodynamics; it allows longer periods of brighter daylight when days are shorter; and, to use an ancient example of dynamic response from indigenous desert architecture, its thermal mass is tuned to cycles of diurnal temperature.

Dynamic buildings have flexible and malleable configurations. There may even be dynamic and movable building components, and their motion might be automated by controls that learn and adapt to use and climate. Switchable window-glazing is one such recent innovation.

The point of this comparison is to establish the systems basis of design and to illustrate how those dynamics invoke complexity. The following subsections will press this issue.

3.1 Dialectic balance

Think of a barren but safe habitat for rabbits where there are two different borders of accessible grasslands for feeding – say, one at each end of their territory. To the south there is abundant food, but also an abundance of predator wolves. To the north there are fewer predators, but the grazing is inadequate to sustain many rabbits for very long. In this scenario we would see that the rabbits move about in patterns that reflect the complex forces of finding food and of avoiding being eaten in turn by the wolves. While each rabbit would essentially act as an independent agent largely irresponsive of the action of others, the pattern of the rabbits' collective behaviour would still be observable as the result of these bimodal force stressors. Some rabbits would subsist on the scrubbier growth of the north range, but most would be forced sometimes to venture south. Interactively, as more rabbits leave for the south, there is more to go around in the sparser north range, and more rabbits will move back in that direction – and so the complex system of migration begins and then continuously cycles through a non-linear and spontaneous, yet ordered and responsive, sequence.

Similar dialectic and generative forces exist as drivers of complex interaction in other realms much closer to building design. In each case there is a pattern of oscillation in complex response to a set of generative tensions, with a resulting non-linear pattern of behaviour that serves the purpose and perpetuation of the system as survival or adaptation. Some relevant applications follow.

- *The ideal organism* (Barrow 1995). The ideal organism oscillates between the epicurean state of calm relaxation and the adventure state of seeking new resources. As the environment varies through the seasons and through the day, the organism will move from safety to risk as appropriate.
- *Natural systems.* The teleologic force of a system keeps it on the edge of chaos in a state of ambiguity between stasis and chaos. A system adapts to uncertainty in response to driving potentials. Like many dialectics, one pole is death by homogenisation of potentials and the other pole is death by randomness and failure of order. A continual or periodic perturbance is needed to keep the dynamic flowing. At one end of the spectrum is noise and chaos; at the other end is stagnation and collapse.
- *Discourse.* A range of active ambiguity is also what keeps discourse at a productive and generative level of energy and motion. Again, when the potential for exchanging ideas is removed, the discourse ends in lack of energy; when the potential meets total disagreement, then discourse ends in outright conflict.
- *Design.* Ambiguity in design rests in the level of complex interaction between the ideal and the real. On the one hand is the ideal, sublime and immediate; while on the other hand is the real, insightful and intelligent.

3.2 Systems not symptoms

To close this section, consider a couple of analogies of systemic versus symptomatic behaviour. First, if you seek treatment for a headache and are offered only aspirin, this is just a symptomatic approach to solving the problem superficially without regard to its systemic cause. Maybe you should have your vision checked to identify a possible underlying cause of the headaches? Likewise, compare the production of lumber from a tree farm versus sustainable harvesting from a forest. The tree farm is an obvious and awkward machine that requires escalating amounts of fertiliser, pesticide and human effort to produce ever decreasing amounts of declining-quality lumber. The forest, on the other hand, is an open system of inter-related parts acting independently, but on the whole produces emergent effects that are quite beautiful. As we extract lumber from the forest at a sustainable rate, it maintains its own level of production without our effort or interference. The forest will also sustain other vital elements of our ecological habitat, and if lightning should strike it down it will grow back much as it was. Not so for the tree farm.

4 The four modes of complexity in building design

Table 2.4 expands on Table 2.3 and offers a framework for discussing four distinct modes of complexity in building design as extensions of complexity science (Bachman 2008). The purposeful end of each mode is summarised as its *Objective*, so that the equivalent of *adaptation* in the scientific sense can be considered in terms of a building design's attributes: embodying human *intelligence*, being authentically *spontaneous*, capturing the teleological *essence* of its origin, and being a *responsive* element within a larger natural context. The connection of complexity in building design with scientific understanding of complexity binds them into a beginning embrace via a common basis of contemporary understanding of the human condition. By relation, then, a tie is also made between building design and its setting in postindustrial social complexity.

This framework of four complexities is offered as a way of considering the relation that building design has with complexity, and how this has been manifest. This is not intended as an authoritative and exhaustive taxonomy, but as a starting point for discussion that other authors will hopefully take up.

4.1 Wicked complexity

The 'wicked problems' definition of design complexity was formulated by Rittel and Webber (1973). They state that design is an indeterminate problem, as distinct from the class of simple problems (see also Buchanan 1995). Tic-Tack-Toe (Noughts and Crosses), for example, is a simple problem because it has a well-defined beginning state (nine-square matrix), operational rules (X or O), and a known end-solution state (three in a row). Design, as Rittel and Webber show, is not simple because it clearly lacks a knowable complete set of beginning conditions owing to the endless amount of information

Table 2.4 A framework of four complexities in building design as compared to the science of complexity (Bachman 2008)

	SCIENTIFIC	WICKED	MESSY	ORDERED	NATURAL
PROPONENTS	Ackoff, Arnhof, Churchman . . .	Simon, Rittel, Bell . . .	Jacobs, Venturi, Alexander . . .	Peña, Sanoff, Preiser . . .	Geddes, Kepes, Olgyay . . .
POSTWAR DATES	1957	1957	1961	1965	1963
REALM	Cosmology	Society	Culture	Institution	Organisms
OBJECTIVE	Adaptation	Intelligence	Spontaneity	Essence	Responsive
ORGANISATION	Dynamic	Cybernetic	Organic	Collaborative	Holistic
ORDER	Interactive	Managed	Authentic	Discovered	Emergent
PROBLEM	Evolve	Bounded cognition	Identity	Reduce data	Flow
AGENTS	Systems	Decision-makers	Citizens	Stakeholders	Systems
APPLICATION	Behaviour	Organisation	Urban	Definition	Design

that could be collected before beginning. Furthermore design has no clear procedural operation of solving the problem or even of addressing its challenges. Finally, design has no one right answer or final point of completion; solutions are bad, good, or better, but never final or perfect; and design simply reaches the end of allocated time, it is never actually finished. While the complete scope of Rittel and Webber's treatment of design wickedness cannot be done justice here, their rendering of the indeterminate and hence complex nature of the design activity in general should be clear.

Beyond the fundamental complexity of design as a human activity and pursuit, there is also the wickedness of designerly thinking as a process. Herbert Simon (1957, 1976) describes our limited cognitive ability as 'bounded rationality' to describe the less than perfect ability to understand a problem completely and work it through to some ultimate solution. Instead, Simon coined the term 'satisficing' to name what problem-solvers in any activity actually do when confronted with vast amounts of information, a finite amount of time, a mixed set of priorities, and a limited set of resources: we always settle for what is 'good enough' given the limitations of what we have to work with and our restricted abilities to think through it all.

To summarise this mode, wicked design complexity is the equivalent of a cybernetic problem in science: a predicament of information feedback and biological control mechanisms. Like the other modes of complexity in building design, wickedness thus takes on the characteristics of a dynamic system and must be treated as such. When we fail to recognise the indeterminate nature of design pursuits and the bounded rationality of human thinking, we not only understate the design challenge, we also misconstrue its true complexity. The consequences of such oversight are that we either attempt to 'tame' the complexity into simple determinate problems, or alternately that we confuse design's richly indeterminate nature with some mystery that can only be approached by ordained enlightened genius and innate talent. As either a simple game or a holy mystery, design as a human pursuit is utterly diminished.

4.2 Messy complexity

Returning to the 1960s emergence of complexity in building design, Robert Venturi has argued for 'messy vitality over obvious unity . . . an architecture of complexity and contradiction . . . It must embody the difficult unity of inclusion rather than the easy unity of exclusion' (1966: 22). Two of Venturi's contemporaries parallel this in their own terms: the urban thinker Jane Jacobs and the architect/mathematician Christopher Alexander. Jacobs recognised that complexity was a special kind of order that stemmed from authenticity and spontaneity:

> There is a quality even meaner than outright ugliness or disorder, and this meaner quality is the dishonest mask of pretended order, achieved by ignoring or suppressing the real order that is struggling to exist and to be served.
>
> (Jacobs 1961: 15)

Alexander's take on messy complexity is friendly to that of Venturi and Jacobs but also craves recognition of a natural human language of form. Note how this statement reinforces the other two writers and also rhymes with the indeterminate character of wicked complexity:

> We are searching for some kind of harmony between two intangibles: a form which we have not yet designed and a context which we cannot properly describe.
>
> (Alexander 1964: 26–7)

Messy complexity thus requires us to deal with the holistic, inclusive and authentic criteria of any design challenge so that we might discover the nascent order that already underlies the problem. Otherwise we are apt to impose a preconceived and inappropriate solution whose form is too wilful in its invention.

4.3 Ordered complexity

In order to 'invent the future', as Buckminster Fuller put it, we must first know where we want to be when we get there. Defining this goal state in building design has been the objective of architectural programming as championed by the likes of Peña (1969), Sanoff (1968), Preiser (1978) and their protégés of today. Although this work is sometimes omitted from discussions on the more noble qualities of design, it is increasingly clear that the strategic goal-setting of design is part of its underlying complex richness. It would be more accurate to think of this ordering sort of complexity as a search for the unique essence – what Aristotle called the 'final cause'. This is what we might hold to be the kernel of the design challenge, the teleological acorn of the oak tree.

In terms of Peña's proactive Problem Seeking, ordered complexity is more than the task of composing programmatic space lists and adjacency diagrams; it is also concerned with collaborative teamwork, multiple conflicting stakeholders, and long-term scenario planning. To use the parallel field of scenario planning as a positive example, the work of ordered complexity in building design is to root out and differentiate a task environment where we can distinguish known immutable determinates, unknowable and irrelevant noisy information, and the region of rich ambiguity where design decisions are most fruitful. Unless we recognise the well-paved avenue of ordered complexity, design is likely to miss the target and wander in unproductive explorations.

4.4 Natural complexity

It is clear that industrial-era practices of design for the built environment have, in the overall scale, spoiled and diminished the vitality, the beauty and even the viability of our natural environment and our built habitat. Those old linear throughput processes of converting resource to buildings and then to used-up scrap now seem short-sighted, greedy, and poorly informed as to our true long-term and global interests. Not only were the buildings and cities themselves becoming increasingly unsustainable, but also the life-support mechanisms of delivering, maintaining and operating them were even worse. In most cases, the most beautiful and useful building could ultimately be associated with ugliness created elsewhere on which it relied for its construction and occupancy. So what once looked liked the road of steady profit and progress is now revealed as a dead-end alley with a sudden cliff beyond; we took the wrong turn somewhere.

As we begin to turn to postindustrial thinking, however, there is much to suggest that the cyclical and complex systems behaviour of natural processes makes a superior model for building design. This fourth mode of complexity in building design can be traced to a constant thread of design-with-nature approaches adopted in all phases of history; but, in keeping with the 1960s theme of postindustrial transformation, the landmarks of interest here would include Victor Olgyay's *Design with Climate* (1963), Ian McHarg's *Design with Nature* (1969) and Rachel Carson's *Silent Spring* (1962).

Since those origins, natural complexity in building design can be said to have developed in five different dimensions – ecology, flow, morphogenesis, synergistics and Gestalt psychology (Bachman 2008). The ecological approach seeks harmony with the energy flows and biological processes of nature as they impact the site and the building. This is closely related to the second dimension as postulated by Norberg-Schulz (1966) as that of flow, where a building can be thought of in terms of reservoirs, conduits, capacitors, filters, barriers, switches, and even transformers of flow. As Steven Groák (1992) pointed out, systems of flow can pertain to light, air, gravity, sound and heat, but they also relate to flows of people, paper, information and product.

The third dimension of natural complexity, that of morphogenesis, is a more direct emulation or even literal adaptation of biological mechanisms into building form. The self-organising, self-emergent, self-regulating and self-reproducing abilities of natural organisms are seen in this dimension as paradigms which design should rather then reinvent in a less systemic and consequently more mechanistic imitation. This is a complex *systems-as-species* approach that regards a building as an organism. As most complex systems we encounter are those of living organisms, it stands to reason that a building which is part of a life-sustaining environment, built or natural, would itself function as an organism and take on the characteristics of living things.

Integration constitutes a fourth dimension of the ecologically complex mode in building design. This concerns holistic, global, macro-scale reality as opposed to local perception of isolated action. It also evokes the integral differences between the whole versus the simple sum of its parts. Buckminster Fuller (1963, 1969) made this part of the architect's language with concepts like 'synergistics', 'tensegrity' and 'empheralisation'. Several more current works look specifically at the integration of building systems as hardware, kits, grammar and species (Guise 1985; Rush 1986; Bovill 1991; Bachman 2003).

Finally, a fifth dimension of ecological complexity is found in a branch of Gestalt psychology with direct associations to art and architecture. György Kepes (1944,

1965, 1966) and Rudolf Arnheim (1954, 1969, 1971, 1977) are most prominent here, especially in the time-line of postindustrial transformation. Their central idea can be described in how the human mind recognises unity and harmony of interacting elements and dynamic relationships. Beyond recognition, there is an aesthetic attraction, significance and satisfaction in these perceptions to which humans are intrinsically drawn. In this appreciation, buildings assume the role of our place in nature, not as artificial space divorced from it . . . except for the utility lines.

5 Four convergences

The four complexities of building design can be situated within today's social context by another set of overlapping frameworks: four convergences we experience in present events. In each case, complexity in building design is facilitated and promoted by the new-found need and innovative ability of the designer to do complex work.

5.1 Bounded rationality meets computer cybernetics

When complex answers become more accessible, the underlying questions become more significant and we are more apt to enquire into the depth of their nature. Computer simulation, database operations, optimisation procedures and other tools of the postindustrial information age have had just that effect. These same tools have also opened up new design possibilities by offering 'what-if' and 'push–pull' explorations that were previously too time-consuming and mind-numbing to be practically delved into. Indeed, there is now an entire literature just on the use of these tools for visualising information that has only recently been made practical to collect and ponder over.

Beyond what the tools avail for us directly, we are entering an era in which design, engineering, production and assembly are increasingly unified by shared computer-model workspaces. This leads to the prospect of mass-customisation in building components; the impact of which will likely exceed the advent of their mass-production. In the mass-customisation scenario, the designer's drawings become the engineer's parametrics. Several iterations and refinements later, the same document becomes the manufacturer's production requirements and the machine instruction set for precision computational driven fabrication. Along the way, the actual design of a component or even of the building form in general might be shaped by algorithms that capture aerodynamic response, minimal structural thicknesses, or other influential flows. The resulting form would be one that embodies a great deal of information and intelligence, and then reproduces the appropriate response with elegant precision, along with the component hardware for assembly.

Along another line, automation of building operation and performance is replacing reliance on over-design for static conditions and one-way automatic controls. In short, buildings are becoming intelligent robots that can interact with building occupants through publicly visible 'dashboards' or genie-in-the-bottle-like avatars of the building smiling up at us from our office computer screen. These prospects for how information animates design and operation of buildings are unbounded and have already been unleashed.

In the aftermath of this convergence of computer cybernetics and bounded human rationality, building design has lost most of its historical insulation from objective accountability. Architecture was described in terms of art alone only for so long as its investigations and productions were considered to be solely the result of inspired genius rather than of deliberate intelligence. Instead, the clinical database of building design is now being built from postoccupancy evaluation, commissioning studies, sustainable design, evidence-based design, and a host of high-performance case studies. Again, full-minded animation requires complex and appropriate interaction of both artful inspiration and intelligent foresight. Anything less is a Frankenstein.

Wicked complexity in building design is the most relevant of the four modes in this convergence, but operations of the structured complexity mode will also benefit as information tools and knowledge society promote productive collaboration with multiple stakeholders and proactive examination of alternative future scenarios within multiple variations of the design master plan.

5.2 Consumptive growth meets environmental limits

A second convergence which ties complexity to current transformations in our society, and hence to building design, is that of environmental, climatic and resource depletion. As 5 billion more members of our population begin to enjoy the industrial prosperity that only 1 billion of us benefit from today, the limits of available resources for us to consume are being stretched. Earth's carrying capacity of the waste pollution from this consumptive growth is deteriorating even faster. Then there is the matter of 1 billion or 2 billion new people to serve as population grows.

This convergence begs for a radical change away from our 'paleotechnic' industrial-age lineage of harvesting, production and pollution. This old model was linear, simplistic and mechanistic. The new 'eotechnic' model is non-hierarchical, complex and deeply inter-related. In short, we are abandoning machine-production practices and opting for holistic approaches that recognise interconnected and dynamic relationships. We finally see that the world is not a machine made for our immediate profit; it is a complex system of adaptation and change in which we can either participate and flourish, or flounder and perish.

This convergence relates directly to the natural mode of complexity in building design. It might also have roots in messy complexity via the aspect of how a building achieves authenticity. Finally, one could also argue that the mode of ordered complexity would identify the essential character of a building, and hence its appropriate ecological niche.

5.3 *Local perception meets global dynamics*

Complex systems theory in science describes the cosmos as an interconnected set of non-linear systems whose cause-and-effect chains can produce non-intuitive outcomes featuring the special order of chaotic patterns. Our evolving world of social complexity reflects this as globalisation reveals the interconnectedness of culture, economy, trade, and behaves increasingly like a set of non-linear dynamic systems. The underlying interdependence of these many geographically and temporally separate events illustrates that our typically local perception of events and their shallow association with direct cause-and-effect outcomes needs to be reconsidered.

Building design must accommodate this convergence of local-scale perception of isolated events and the new realisation of their macro-scale inter-relatedness. The immediate perception of a building's apparent 'commodity, firmness, and delight' needs to be supplemented with an evaluation of its long-term and far-reaching impacts. As this recognition takes hold, we already see tangible examples of its impact on building design practice: postoccupancy evaluation, continuous commissioning and, of course, sustainable design, just to mention a few.

As a convergence of social trends and building design practice, this idea is relevant to a range of the complexity modes: natural complexity and sustainability, ordered complexity and postoccupancy, messy complexity and inter-relatedness, and wicked complexity of dynamic systems.

5.4 *Cost meets long-term value*

A cynic, as Oscar Wilde tells us, is someone 'who knows the price of everything and the value of nothing' (Wilde 1892). As a postindustrial notion, we see this idea of cost versus value gaining currency in attitudes favouring long-term investment rather than short-term profit. This reflects the postindustrial principle that value is produced by knowledge workers employing the intelligent collection, interpretation and inference of information. Building design is caught up in this, too: strategic planning, life-cycle economics, organisational psychology, learning organisations, value engineering, and sustainable design are frequent examples. The fundamental realisation that building construction costs and operational expenses are almost insignificant compared to their monetised impact on occupant satisfaction or productivity is also a long-overlooked aspect of this.

Like the other three convergences of society and complexity, the new emphasis on value over cost and investment over quick profit is a stimulus for complexity in building design. The modes of wicked, messy, ordered and natural are all supported by taking a long-term approach to value in lieu of a short-term approach to cost.

6 Embracing complexity

Postindustrial transformations have served in this discussion to emphasise the increasingly complex nature of buildings and of building design. And, while the convergence of society and complexity in our times seems to begin in the 1960s, it is also clear that this complexity has always existed in some measure. Having covered that, we now turn to the central question of this chapter: How do we now authentically embrace complexity in building design rather than seeking to master it falsely or reductively tame it into sets of simple problems? Four possibilities can be suggested:

6.1 Animation

Consider the human brain and mind as a simplified but comprehensible analogy for complexity in design. The brain is nothing more than an organ like a lung or a kidney; the mind, however, is something quite different that emerges out of complex interactions within the brain's neural network, especially between its rational left hemisphere and its emotive right hemisphere as facilitated by the mediating membrane of the *corpus callosum*. In the end, however, the mind is self-emergent and self-ordering; it does not even reside in the brain (Kelso 1995). The self-emergent mind is part of the animation of life, the difference between the whole and the sum of its parts. It is the complexity.

Frankenstein's monster is a negative example of this animation. Like all monsters, it was not truly alive in the manner of a living system. The monster is always just a machine imitating the complexity of life. While Mary Shelley (1818) may have written *Frankenstein* as a cautionary tale against the looming inhumanities of the dawning industrial smokestack age, this is also a fair comparison to buildings. On the one hand, we have buildings that successfully resolve conflicting needs, ideas, values and priorities; they address all four modes of complexity and attain fully mindful animation as dynamic systems. On the other hand, we have buildings which respond only to simple criteria and symptomatic evaluation. Again and again these later buildings, however pleasant they seem at first in the local scale of perception, will ultimately become Frankenstein's monster when macro-scale inter-relation is at the castle gate.

If, as science and literature seem to claim, complexity is essential to full-minded animation, then building design must embrace all modes of this complexity. There is no recipe for such a pursuit, but the first step is to decide on systemic-level evaluation rather than symptomatic ones.

6.2 Unique essence

The teleological relation of the acorn to the oak tree should be understood in building design as a search for a singular unique essence to which design can respond. Without first identifying this essence, design work will wander, guess and assume rather than understand, structure and respond. While the more inventive side of human intention is vital to the translation of the essence into built form, that teleological essence is first required. Within this exchange of intentionality versus responsiveness the designer must balance the influences of discovered form versus invented form. As discussed earlier in this chapter, and as taken historically from Peña's *Problem Seeking* forward, that balance is presently shifting away from the more wilful and stylistic approaches on design and moving towards a path of responsive enquiry into the nascent aspects of project criteria. Sustainability, strategic services, scenario planning, and postoccupancy evaluation are just a few such examples already presented. A more in-depth account of this discovered or revealed form would also invoke a philosophical background including Heidegger's existentialism and Schopenhauer's 'will existence'. In the postindustrial framework offered in this writing, however, the modes of complexity as already discussed can be embraced by recognising design statements in their complex inclusive whole of mismatched requirements, resources, conditions and expectations. Within that inclusive description lies the acorn of unique essence and the rich ambiguity surrounding it where design can operate so productively.

6.3 Hermeneutics

A more prescriptive approach to embracing complexity is also possible. This third posture of embrace is vested in a cognitive framework for designerly thinking that is neither bottom-up deductive nor top-down inductive; but rather abductive in the sense of a cycle of beginning observation, then learning, proposing, testing of fit, and then beginning again with a new understanding of the fundamental problem statement. Fortunately this abductive thinking is a process with which designers are already patently familiar as an approach to complicated levels of information on the criteria and programming of a project. Elevating this from a means of dealing with complicated information to one of managing complex relations begins by recognising the propositional basis of most innovation. It also requires a critically questioning attitude of making the abductive process more explicit, rigorous, and deliberate rather than the frequently self-conscious, impressionistic and personally gratifying pursuits. Here, again, the typically local scale of everyday superficial perception needs to be supplemented by broader thinking across macro- and micro-scale relations. Thereafter the abductive design exercise can be made holistic and inclusive rather than selective and intermittent through cycles of hermeneutic learning and interpretation.

Hermeneutics describes that iterative cycle of abduction – from observation to learning, to proposition, to testing, and through redefinition of the problem back to observation again. As a branch of contemporary philosophy and science, it is also a good method for deploying the limits of bounded human rationality against the wicked complexity of building design. Because the process is iterative, it can be thought of as a constant circular spiralling down towards a final solution point at which the unique essence of the design problem waits like a strange attractor. We can also imagine that there is a pair of dialectic forces at play in propelling the design work: a centripetal inward pull and a centrifugal outward tug. Each of these forces is powered by a persistent questioning attitude in design: 'Why not?' is the outward tug of force that always expands possibility in the opportunistic quest for better and richer alternatives; 'So what?', on the other hand, is the contractive force that draws the design gradually towards completion. As suggested by the spiral path, there is no straight-line, direct and immediate connection between the wicked problem and its complex solution. Rather, a constant questioning and continuous cycle of learning are required, as well as a willingness continually to re-conceptualise the problem statement.

6.4 Aesthetics

A fourth mode of embracing complexity is that of aesthetics, which is defined for the purposes discussed here as the connection between our understanding of something and our appreciation of it. We can also define aesthetics in this sense as how design is employed to connect the ideal with the real – or, in other words, the search for a beautiful solution with the realisation of a construction. As the formal meaning of aesthetics involves the philosophy and study of human appreciation of beauty, these special definitions are aligned with its accepted normal usage.

Embracing complexity thus affords the aesthetic pursuit of connecting the idealised 'unique essence' of a design problem with the reality of its solution and manifestation. In the process it is clear that the designer will connect his understanding of the design with his appreciation of it. If the built solution is true to these principles, there is a significant likelihood that the building will become an authentic part of its setting, neighbourhood, and other scales of relation.

7 Conclusions

This discussion shows how building design involves not only complexity of information detail but also a much deeper and richer dynamic complexity. It is in this attitude that the systems basis of building design can find a compelling rationale as well as a cohesive means of embracing complexity. Chronology shows that, while the complexity has always existed in this realm, the history of its embrace in building design parallels our transformation into a postindustrial knowledge society and reflects similar transformations in many professions.

Reasoning frames three sets of operations that are presented here in sequential lists. First, four modes of the encounter with complexity in building design are traced as a framework for exploring its manifestations. Next, four convergences of historical trends in our present lives are shown to promote a complex dynamic systems view of buildings and building design. Finally, four embraces are offered as designerly ways of thought which take complexity as the rich ambiguity and true unique essence of a building proposition.

Complexity in building design is of course set against some prevailing attitudes; but in many cases the opposition lies merely in making the embrace of complexity an explicit and accountable process rather than a purely intuitive or normative one. What is required for reconciliation is a harmony of invented versus discovered form, and a recognition of the macro- and micro-scale context of design beyond its local effects on our immediate perception.

Acknowledgements

This chapter contribution is part of an on-going project towards my own book on what I term as 'strategic design' as the complimentary adjunct to physical design. To date, the book project has been piloted as a series of papers, each exploring one of the project's central propositions. Early on in this investigation, I was especially privileged to have been invited into the CIB workgroup on Embracing Complexity in Building Design and to have participated in their 2007 ECID Workshop in Liverpool. I met a great many people there who I am now happy to deem as colleagues. Special thanks go to Dr Halim Boussabaine, who chaired the event and extended my invitation.

References

Alexander, C. (1964), *Notes on the Synthesis of Form*, Cambridge, Mass.: Harvard University Press.

Arnheim, R. (1954), *Art and Visual Perception: A Psychology of the Creative Eye*, Berkeley, Calif.: University of California Press.

Arnheim, R. (1969), *Visual Thinking*, Berkeley, Calif.: University of California Press.

Arnheim, R. (1971), *Entropy and Art: An Essay on Disorder and Order*, Berkeley, Calif.: University of California Press.

Arnheim, R. (1977), *The Dynamics of Architectural Form: Based on the 1975 Mary Duke Biddle Lectures at the Cooper Union*, Berkeley, Calif.: University of California Press.

Bachman, L. R. (2003), *Integrated Buildings: The Systems Basis of Architecture*, New York: John Wiley.

Bachman, L. R. (2006), 'Postindustrial architecture, dynamic complexity, and the emerging principles of strategic design, *Proceedings of the ARCC/EAAE International Conference on Architectural Research*. The Architectural Research Centers Consortium. Available on the Internet at www.arccweb.org

Bachman, L. R. (2008), 'Architecture and the four encounters with complexity', *Architectural Engineering and Design Management*, 4: 15–30.

Barrow, J. D. (1995), *The Artful Universe*, Oxford: Clarendon Press.

Bovill, C. (1991), *Architectural Design: Integration of Structural and Environmental Systems*, New York: Van Nostrand Reinhold.

Buchanan, R. (1995), *Discovering Design: Explorations in Design Studies*, Chicago, Ill.: University of Chicago Press.

Carson, R. (1962), *Silent Spring*, Boston, Mass.: Houghton Mifflin.

Cherry, E. (1999), *Programming for Design: From Theory to Practice*, New York: John Wiley.

Fuller, R. B. (1963), *Ideas and Integrities: A Spontaneous Autobiographical Disclosure*, Englewood Cliffs, NJ: Prentice-Hall.

Fuller, R. B. (1969), *Utopia or Oblivion: The Prospects for Humanity*, New York: Bantam Books.

Geddes, P. (1904), *City Development: A Study of Parks, Gardens and Culture-institutes. A Report to the Carnegie Dunfermline Trust*, Edinburgh: Geddes & Co.

Groák, S. (1992), *The Idea of Building: Thought and Action in the Design and Production of Buildings*, London: Spon.

Guise, D. (1985), *Design and Technology in Architecture*, New York: John Wiley.

Jacobs, J. (1961), *The Death and Life of Great American Cities*, New York: Random House.

Kelso, J. A. S. (1995), *Dynamic Patterns: The Self-Organization of Brain and Behavior*, Cambridge, Mass.: MIT Press.

Kepes, G. (1944), *Language of Vision*, Chicago, Ill.: Theobald.

Kepes, G. (1965), *Structure in Art and in Science*, New York: Braziller.

Kepes, G. (1966), *The Man-Made Object*, New York: Braziller.

Lyle, J. T. (1994), *Regenerative Design for Sustainable Development*, New York: John Wiley.

McHarg, I. (1969), *Design with Nature*, Garden City, NY: Natural History Press.

Mumford, L. (1922), *The Story of Utopias*, New York: Boni & Liveright.

Norberg-Schulz, C. (1966), *Intentions in Architecture*, Cambridge, Mass.: MIT Press.

Olgyay, V. (1963), *Design with Climate: Bioclimatic Approach to Architectural Regionalism*, Princeton, NJ: Princeton University Press.

Peña, W. M. (1969), *Problem Seeking: New Directions in Architectural Programming*, Houston, Tex.: Caudill Rowlett Scott.

Preiser, W. F. E. (1978), *Facility Programming: Methods and Applications*, Stroudsburg, Pa: Dowden, Hutchinson & Ross.

Rittel, H., and Webber, M. (1973), 'Dilemmas in a general theory of planning', *Policy Sciences*, 4: 155–69.

Rush, R. D. (ed.) (1986), *The Building Systems Integration Handbook*, New York: John Wiley.

Sanoff, H. (1968), *Techniques of Evaluation for Designers*, Raleigh, NC: North Carolina State University, Design Research Laboratory, School of Design.

Shelley, Mary (1818), *Frankenstein*, London: Lackington, Hughes, Harding, Mavor & Jones.

Simon, H. A. (1957), *Models of Man: Social and Rational; Mathematical Essays on Rational Human Behavior in Society Setting*, New York: John Wiley.

Simon, H. A. (1969), *The Sciences of the Artificial*, Cambridge, Mass.: MIT Press.

Simon, H. A. (1976), *Administrative Behavior: A Study of Decision-Making Processes in Administrative Organization*, New York: Free Press.

Venturi, R. (1966), *Complexity and Contradiction in Architecture*, New York: Museum of Modern Art, distributed by Doubleday.

Wilde, O. (1892), *Lady Windermere's Fan*.

3 Complexity in engineering design

Claudia Eckert, René Keller and John Clarkson

1 Introduction

'Cars are complex, trains are complex, aeroplanes are complex.' Very few people would disagree with this statement, but many would not be able to express the nature of this intuitively understood complexity. Not just laypeople but also professional engineers experience these products as complex. Even as highly trained experts they can only understand a small part of the product and the system within which it is deployed and operates. They do not reflect about exactly what complexity means for their products and how they are affected by complexity, but they marvel at how the effort of a very large team can bring forth a product which has such tightly defined behaviour, which is consistently delivered under very varied circumstances.

The sheer scale of large engineering projects is impressive in terms of the number of people involved, the number of components designed and the time-scale of the project. For example, an aircraft has tens of thousands of components designed and developed by teams of thousands of engineers at the main aircraft manufacturer in conjunction with a large supply chain over a period of about five to ten years. Major components such as the avionics, the undercarriage or the engine are sourced to main first-tier suppliers. These major components themselves are highly complex products with hundreds of components designed by a large team of people and long supply chains. For example, designers have commented in discussions that the design effort going into a jet engine is about 15 million person hours.

To give an impression of the factors that affect the complexity of engineering designs, this chapter will use jet engines as an example to illustrate how different types of complexity manifest themselves in engineering design. The focus of this chapter is on engineering changes, which are an important part of design work – as small changes to certain parts of the engine can result in either much better performance or major rework necessary to fight disastrous effects of this change – as it is impossible to do justice to all the aspects of complexity in engineering. Therefore, section 2 will focus on general theories of complexity and the interplay of the factors affecting engineering design, leading to section 3 which will look in more detail into engineering change as a driver of complexity. Section 4 will introduce a method that has been developed to reduce some of the complexity of engineering changes by providing risk measures, visualisation and decision support. How, in particular, this method is addressing complexity issues regarding complex engineering products (i.e. jet engines) is the subject of section 5.

2 Types of complexity in engineering: example of a jet engine

Jet engines themselves are highly complex products in many different ways. The engines are highly optimised to maximise fuel efficiency and minimise cost and risk. For many of their components, the performance optimum lies close to a cliff edge, after which the behaviour deteriorates into chaos (Alligood *et al.* 2001), in the sense of bounded unpredictability (i.e. it is known what kind of behaviour the system will display, but not exactly how it will behave). Therefore small changes to key components can have potentially disastrous effects and affect many parts of the entire system, as components are pushed beyond the optimal point to rapidly deteriorating behaviour. Components are often designed with a certain degree of redundancy or margin in them if it is expected that many further changes might be required. A simple example would be that turbine blades are designed with a certain clearance to their outer casing. If the turbine blade were made longer beyond a certain point, the casing would need to be redesigned. Several small changes occurring in parallel can also push components over the edge and lead to major consequent changes. Spotting and avoiding the changes is a great management challenge, which requires a very good understanding of the product and the processes associated with it.

However, very few people have an overview of the product. The difficulty and degree of specialisation is such that engineers estimate that it takes fifteen years to become an expert in a particular aspect of jet engine design. Some engineers specialise in a small detail of the design: for example, one specialist designs the cooling holes of turbine blades and spends his working life identifying the exact shape, location and angle of these very small cooling holes; others integrate different systems. The understanding of the engineers who design the system and the properties of the system co-evolve together. Requirements on the designs and curiosity drive research which shapes the product. The understanding and the technical properties together form an adaptive system, which changes its connectivities and dynamic behaviour in response to its environment, while co-evolving systems develop mutual changes of structure and behaviour (Kauffman and Macready 1995).

Many complex engineering products tend to be big and often hot, and often only operate in sometimes hostile environments – as jet engines do. Therefore engineers cannot directly interact with the system; they have to interact with descriptions of the product. Simon (1998) considers the complex engineered or 'artificial' systems as almost decomposable, that is they are hierarchical to some extent, but not fully decomposed into separate, independent parts. Many components are part of different systems and contribute to different functions; for example, in some engines the fuel is used to cool the engine, linking the fuel system with the cooling system and the details of the mechanical design of the components that need to be cooled. The underlying connectivity representing how the different parts relate to each other determines the constraints and potential for behavior. Connectivities of a complex design form a lattice structure rather than a tree structure, although the latter is often an adequate approximation for almost decomposable systems.

Here the actions of the human actors are played out against the background of the properties of the product, the organisation and the wider market context. The properties of the product and in particular the structure of the product constrain the actions that can be carried out with the product. Engineers can only design what physics allows them but can be produced economically and with what constitutes a sufficiently incremental design to minimise the risk. Over a long period of time the product and the organisation around it are co-evolving. However, looked at from different time-scales, part of the system forms a stable background for others to operate on. In the short term, designers interact with a fairly stable product, which they refine, test or maintain. Over a longer time the actions of the humans remain fairly constant and the product develops incrementally.

Figure 3.1 summarises the different types of complexity in engineering design as discussed in Earl *et al.* (2005). Human actions interact with the background of a system through descriptions, but different theories of complexity set their emphasis in different aspects of this complex system.

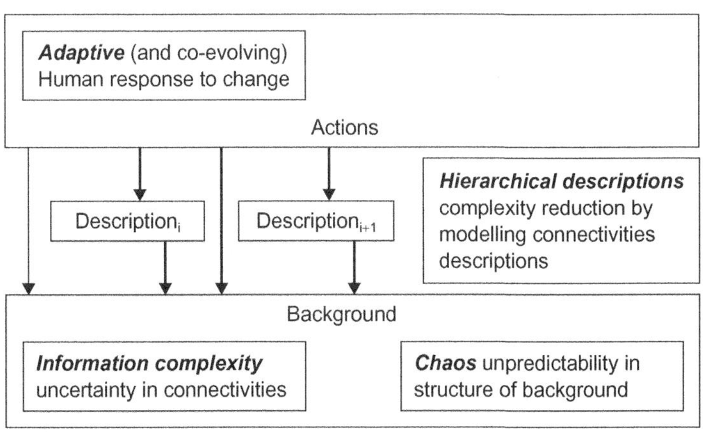

Figure 3.1 Types of complexity most appropriate at each level. Source: Earl *et al.* 2005

2.1 The multiplicity of factors that make engineering design complex

The example of jet engines has illustrated some of the scale and complexity of engineering design problems. Figure 3.2 shows another view of the many interplaying factors affecting engineering design. Most simply, designers generate a product for users through a process. All these elements constrain and affect engineering design and add to the complexity of the activity. This section aims to give a flavour of the influencing factors rather than provide an exhaustive discussion of the sources of complexity.

2.1.1 User

Users have their own requirements and preferences, some of which they can express explicitly, such as 'the flight range', while others are hard to quantify, like 'quality of noise'. Each user has his own sets of requirements and specifications that meet the use profile he has in mind. For example, an engine is selected according to the intended use profile, and the aftercare packages sold with the engine are tailored to the particular intended use.

Engineering is heavily regulated, so that many requirements do not come from the user directly but from the regulatory environment within which the engine will operate. For aviation, these regulations are international; but, for example, for diesel engines, these are set by individual countries, where some set very stringent requirements and others have no regulation at all, but are likely to set regulation soon (Jarratt *et al.* 2003). At the moment engine-makers need to estimate what will be the regulatory requirement of the huge Chinese and Russian markets, when these come into being during the life-cycle of the current or next generation of products.

In the case of engines, regulation and legal requirement are a reflection of environmental concerns and needs. Emission regulations protect the environment by

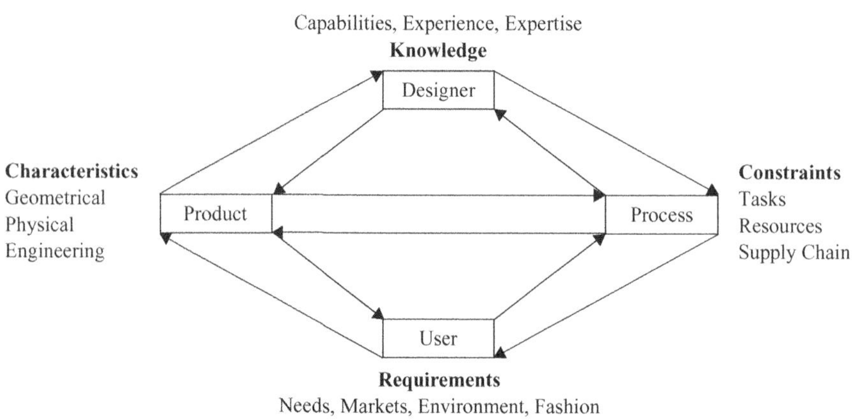

Figure 3.2 Interplaying factors in engineering design

explicitly setting limits on what can be sold. Issues like fuel consumption are not set explicitly but driven by market demands, as customers select the most fuel-efficient options. The values of soft issues like noise are often socially negotiated, and companies have to understand and anticipate what will be required over the life-time of the product.

Engines are not fashion products, but other products like cars are subject to changes in fashion, which will affect the customers' intuitive response to the styling of the car. This is extremely volatile and hard to predict, but has a huge impact on the profitability of the product. The sale of products is affected by the competitor products on offer, from which the product must distinguish itself. Many products, like the famous example of the Ford Ethel, were not bad designs in their own right but failed because the markets had moved on or competitor products had a clear competitive advantage. A more contemporary example is the increasing difficulty in selling SUV on the US markets, as fuel prices are going up and disposable income is going down.

2.1.2 Designer

Designers have to translate these requirements into products. Each person brings his own skills and knowledge to the design, as well as his personal experiences and preferences. Every design is fundamentally influenced by the people who generate it. As most engineering projects are carried out by large teams, the interaction between the different people and the dynamics in different teams varies. Design teams are also embedded in wider organisations with many business and support functions, as well as in some cases the production organisation. For complex engineering products, this is repeated along a large supply chain. Typically designers do not interact directly with their users or their customers, but this interaction is mediated by a larger organisation, often a sales or marketing department. Along these chains and across these interactions, design specifications, requirements and intents are frequently translated and reinterpreted, leading to iteration and sub-optimality. As a socio-technical process, engineering also has the complexities any social process has in terms of personal preferences, likes, dislikes, and issues associated with any type of collaboration. Most complex engineering products are now produced by international supply chains, which requires the collaboration of people from different fields or, as Bucciarelli calls them, object worlds (Bucciarelli 1996) and different cultures. For example, many jet engines are used in Airbus aircraft, which are designed by cross-European teams, consisting of teams in France, Germany, the United Kingdom and Spain. Each country has its own language and engineering tradition. Traditionally jet engines were designed by mechanical engineers; but now control software, developed by electrical engineers or software engineers, plays a very significant role. Owing to the non-physical nature of control engineering, the processes and work culture are very different from those of mechanical engineering, with people thinking in fundamentally different ways and using different representations and processes.

2.1.3 Processes

In designing, designers and organisations follow processes, in two quite distinct meanings of the word: the actual set of activities, which are carried out to generate a product, and the high-level prescribed set of steps that must be followed to carry out a repeatable task or to generate a product.

Formal or even informal design processes are fundamentally a means of structuring a task which is too large to be carried out without structure. Sequences of activities need to be established and coordinated. Process complexity is therefore directly linked to product complexity. The more complex the product, the more tasks will be associated with designing it, which in turn requires more people or longer processes. As the time to market date is often enforced, most processes are therefore carried out in a parallel fashion. The coordination of these activities adds additional tasks, so that the relationship between product complexity and process complexity is likely not to be linear. In our case studies we noticed a marked change in complexity of the process required to design a diesel engine and that of a jet engine, because a diesel engine is sufficiently simple to be understood by experienced engineers, which is no longer the case with a jet engine (Flanagan *et al.* 2006).

Most engineering design companies now employ a gateway process (Cooper 1993), which sets major milestones and provides checklists for what must be achieved at each milestone. In addition companies have set processes to carry out particular repeated tasks, such as a Computational Fluid Dynamics analysis (CFD), which is required for all flows in a product like a jet engine, or failure mode analyses (Hirsch 2007), which are carried out at all significant decision-making points.

The success of processes is influenced by the inherent difficulty of the design tasks and the availability and quality of the resources that are available to carry it out. These processes are intertwined – horizontally – across multiple projects being carried out in parallel in the same organisation and – vertically – down the supply chain. This intertwining leads to a competition across resources and work shares, and can propagate problems across a network of unconnected project. This is enhanced by a prevailing culture of fire-fighting, in which problems are often dealt with in an 'all hands on deck' mentality, which leads to resources shortages and problems elsewhere.

2.1.4 Product

The product itself has specific characteristics that influence the way it must be designed. While many engineering products do not pose great technical challenges, others, such as almost each new jet engine, push the envelop of what is technically possible. As explained above, the optimal performance of a jet engine lies close to the point of decline in a jet engine. The challenge often lies in designing a product that will operate equally reliably in all possible scenarios of use. The difficulty therein lies in extreme cases of use: for example, jet engines that can operate at high altitude (landing in Lima), in extreme cold (North Canada) as well as in powdery sand (Saudi Arabia). Robustness, reliability and resilience are tacit requirements for most products. Detailed properties of the design can cause huge challenges or difficulties in design processes.

None of these factors can be looked at in isolation. They are all connected and influence each other. Rarely is it possible to assign a clear direction of causality to the analysis of influences and problems, and how these influence work and can be managed and mitigated is a subject of active research in engineering design. Examples for this can be found in many standard engineering design textbooks, such as Ullrich and Eppinger (2004) or Pahl and Beitz (1996).

2.1.5 The relationships

The complexity of engineering design does not lie just in the user, the designer, the processes and the product themselves, but in how they are related. It lies in what users want of the product and in how designers carry out the processes. All these factors are closely related, but not all connections are always visible or developed. For example, many designers have very little contact with users and their needs. The needs are filtered through technical requirements that products have to meet. However, it is rarely that simple. In understanding how exactly the requirements should be met, designers need to have subtle and often tacit understanding of how the customers will interact with the product over its entire life-cycle. They need to guess customers' needs that the customers themselves have not even expressed. This is becoming ever more apparent as products and services carried out with the product are designed together. Product Service Systems PSS (Mont 2002) advocates designing the product and the service as equal partners, i.e. designing the product and the product relationship with the user at the same time. In many very complex engineering products, like a jet engine, the technical challenges in designing the product are sufficiently high that so far the services have been secondary to the product. However, as companies have begun to sell capabilities rather than products, the integration of products and processes has become more prominent. For example, jet engines companies now offer contracts whereby they own the engines, are responsible for the service and maintenance, and their customers pay them per flight hour (Harrison 2006).

The relationship between product and processes depends fundamentally on the interactions of the designers involved. How exactly products and processes affect each other is a topic of on-going research, but the general issues are obvious. Experts are likely to make fewer mistakes and therefore reduce the need for iteration in the process. Simpler or more elegant solutions might take longer to create, but later be much simpler to produce or maintain. Clear processes can support the communication and decision-making, and will increase both job satisfaction and product quality. Many more of these simple arguments can be constructed. These are not rules that will always apply; rather they are patterns, but can frequently occur. The greatest variable in the relationship between products and processes is the amount of new design that is required in a new design. Most complex products are designed by modifications from existing products. These increments can be very small to meet new needs or can be a huge change including significant innovation.

In the remainder of this chapter we shall look at complexity and engineering change and some tools the authors have developed to help designers to cope with complexity. Change processes are typically simpler than design from scratch, but display many of the same characteristics.

3 Complexity and engineering change

Complexity is everything everywhere in engineering design – in the product, the process of generating it and the context in which the design is created, produced and sold. But in particular it lies in the interconnection of these different factors and the subtle influences they have on each other. Many of these influences are ill understood and therefore extremely difficult to describe and model. One area where all these factors come together, but in a slightly simplified form, is in engineering change: modifications made to existing products. Most products are designed by modification from existing designs (Otto and Wood 2001), and designs are modified throughout the entire design process and most of the life-cycle of the product. The remainder of this section uses engineering changes as an example to show how complexity can be interpreted, modelled and reduced in engineering design.

3.1 Engineering change

Very few engineering projects ever start totally from scratch; rather they are based on the company's own past designs or products that are already on the market. Even in very innovative designs, parts of the system remain constant. For example, the Dyson vacuum cleaner is often cited as an example of a highly innovative design (Julier 2000). However, while the suction mechanism did constitute a radical innovation, the vacuum cleaner still had a motor, brushes, cables, etc., just as the vacuum cleaners that came before. Most designs are created through incremental change (McMahon 1994). In some industry sectors, such as the aerospace industry, this is the normal way of designing (Vincenti 1990).

Jarratt (2004) defines an engineering change as '*an alteration made to parts, drawings or software that have already been released during the product design process*'. In incremental design this even stands at the beginning of the design process. Changes occur during the product development processes due to problems occurring with the design or changing requirement – see Eckert *et al.* (2004) for a discussion of the causes of change. Once a product is released, changes often lead to new versions which are generated for particular customers or markets. Later in the life-cycles of some products, the product is upgraded to meet new needs or to be brought in line with newer similar products.

The cost of implementing changes differs, depending on when in the life-cycle of the product they happen. Clark and Fujimoto (1991) suggested a 'rule of 10', whereby the cost of a design change grows by the factor of ten with each passing design phase while the number of changes decreases. The later a change request is raised, the more important it is to contain changes locally. Eckert *et al.* (2007) argue that during the life-cycle of a product the nature of changes and the response to changes evolve. While the bulk of changes happen in the design phases, some lead to redesign within the same product, others lead to versions of the product, while changes within manufacturing and operation either require upgrades to the product or trigger a new design process of a product variant (see Figure 3.3).

Empirical studies of change processes (Eckert *et al.* 2004) have established a distinction between two kinds of changes: initiated changes which are caused by external factors such as legislation, or customer requests and emergent changes which arise from problems in the design or use. One particular kind of emergent changes is knock-on or propagation changes caused by changing other components. Predicting these changes is a major problem in design processes – see Brown (2006) and Clarkson *et al.* (2004). However, regardless of when these changes occur or for what reason, the processes to implement a change are very similar (Eckert *et al.* 2004).

3.2 Interacting with change

Change issues are only one of many aspects of complexity in engineering design, but change is a fundamental part of all design processes and needs to be actively managed. Complexity, like change, cannot be avoided, but it can be managed and to some extent reduced. The problem can be approached from many different angles.

Research that is aimed at supporting designers in handling change focuses on three main aspects, which are also typical for other areas of support for engineering design:

- *Processes* to carry out changes. Maul *et al.* (1992) describe a generic engineering change process as a five-stage process consisting of (1) filtration of change request, (2) development of solution, (3) assessment of solution, (4) authorisation of change and (5) implementation. Many companies have similar processes and put effort into making sure that all steps are carried out swiftly and well.

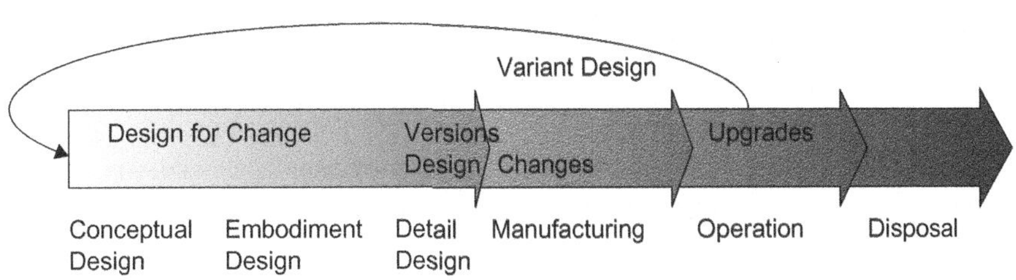

Figure 3.3 Variant design triggers a new design process. Source: Keller *et al.* 2008b

- *Workflow-support tools*, or '*paperwork support tools*' as Jarratt (2004) calls them, focus on how to manage the entire engineering change process efficiently. They support designers in a reactive way, in making sure that the correct information is available at the right time, all stakeholders are informed about change decisions, or information on changes is stored at an accessible place.
- *Decision-support tools*, or simulation stools, which help designers to assess the impact of change or model the consequences of a specific change.

While the previous list is generally applicable to design processes, Fricke *et al.* (2000) propose five attributes to implement a better change management process:

Less: Preventing changes can be achieved by a more 'in depth' analysis before designing and implementing the changes.
Earlier: Considering 'the rule of ten' (Clark and Fujimoto 1991), the earlier a change is implemented into a design the less costly it is (front-loading).
Effective: Deubzer *et al.* (2005) found that 39 per cent of all changes that happen in the later stages of the design process are considered to be avoidable. A better assessment of these changes and potential side-effects – using decision-support tools – should avoid such changes that are not necessary.
Efficient: This strategy aims at improving the change processes. Having an effective change process in place, which makes best use of all available resources, can result in a seamless integration of changes into the general design process.
Better: Learning from previous change cases and change processes will help to improve a company's ability to manage changes in the future.

It is also possible to approach engineering change in product development itself. A product can be designed with a huge redundancy, so that it can accommodate new functionality without requiring it to be changed. This makes a product resilient to change. In many cases this is not possible, because it would increase the production and running cost of the product beyond what is reasonable. A related approach is to manage customer expectations. For example, it pays companies to convince their customers that they want a product that already exists, rather than have one designed for them. In that case a customer might end up with an over-designed but cheaper product for their own purpose.

Changes can also be managed through the product architecture chosen for the product. A modular architecture might enable the company to change a component, but keep its interfaces constant, so that no further changes are required. Many complex products, such as diesel engines, are offered with different modular option packages, which allows the company to exchange components without any redesign. For example, a structural sump can be used instead of a non-structural sump. Many companies achieve economy of scale in their product through using product platforms. In particular, in the car industry components are standardised across several product families and brands. Platform components (Simpson *et al.* 2006) are very rarely changed, because their changes would propagate across several product families.

De Weck and Suh (2006) proposed to design product platform specially in a flexible manner to accommodate more changes. Investing in platform shifts complexity of the product offering to the development process, where significantly greater effort needs to be invested to make sure to satisfy many different product families.

In the design of new products, axiomatic design (Suh 2001) suggests that function and the physical embodiment of the design are created together in such a way that components do not carry out several functions through the same features. This can make products conceptually far simpler, but also potentially heavier and more expensive. Axiomatic design is one of few design strategies that is explicitly developed to reduce complexity in product. It recognises that products carry legacies from starting redesign, and require development from scratch, leading to potentially much larger design effort during the design process. Again, this approach trades product complexity against process complexity.

3.3 Change propagation

One particular type of change is changes that arise from other changes to the product through change propagation. Fundamentally these effects of engineering change are not deterministic but probabilistic. Whether and how a change will propagate depends exactly on how the change is carried out and when in the design process it happens. The structure of the product itself forms the background for the changes since changes propagate through components that are connected, but the exact nature of the link also affects how a change can propagate. As illustrated in Figure 3.4, different types of parameters can link components. The particularly tricky changes to predict are those where the nature of the link changes. For example, if a starting component is made larger, another component might have to shift, so that it gets closer to its neighbour, so that if it gets hot during operation and thus expands it touches the neighbouring component and causes vibration. Here a geometric link has changed nature to a thermal and later a vibration link.

In addition, components are connected through global product parameters, such as weight or cost, which affect all components in a product, and can require changes

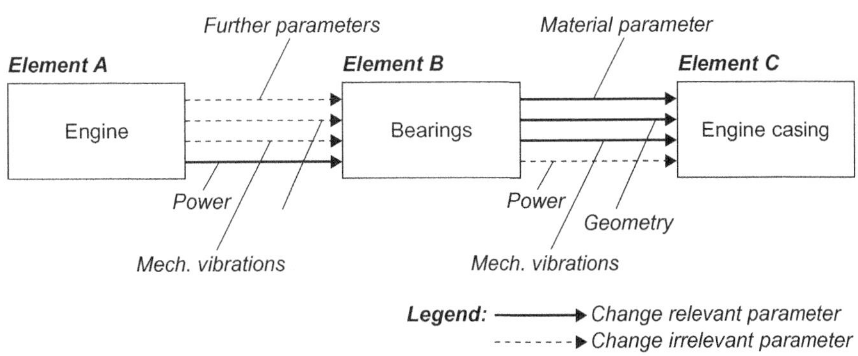

Figure 3.4 Different linkages between parameters. Source: Eckert *et al.* 2004

to otherwise unconnected components. For example, if a component has become a lot heavier through a change, e.g. because a beam needed to be reinforced, other heavy components need to get lighter to maintain the overall weight of a product. These connections are particularly hard to model, because they connect all components, and whether they become relevant depends on the exact circumstances. Whether a change propagates depends fundamentally on whether the component that needs to be changed can absorb this change, or passes the change on to another component, multiplying the change, passing it on to many more components. See Figure 3.5 on the left. For example, a beam can accommodate a certain amount of extra weight, but if it exceeds a certain point the beam has to be reinforced, which is likely to require different fixings, and possibly the wall it is placed into needs to be reinforced, which in turn might need stronger foundations, etc. This depends on the stage of margins on a component that are eroded during the design and use process. For example, the load-bearing requirements on the beam might have been increased several times already before a potentially critical change occurs. Change multipliers are likely to result in a much larger than predicted effort in the change processes, as illustrated in Figure 3.5 on the right. Some change processes ripple out, with little unexpected additional effort, while other changes blossom into a real mini design project. However, when changes multiply in unexpected ways, change processes can go out of control in so-called change avalanches. Sometimes it is possible to bring this to a close with additional resources, but in other cases it might not be possible to carry out the change without proper redesign. This is an example of a product–process link.

If a design can be parameterised, it is possible to analyse which component would be affected by a change. Many CAD systems can analyse the impact of a change; however, they cannot anticipate how designers would respond to a need for change. Designers in most cases have a choice how they respond to a change. They can reject it or respond in a different way. They might not carry out their own task as well as they should, and in consequence further changes might be required. How well and how quickly people can respond to changes is influenced by how familiar they are with the design. During the design process, changes are carried out by the original designers of the components, who are still familiar with the design. Once a product

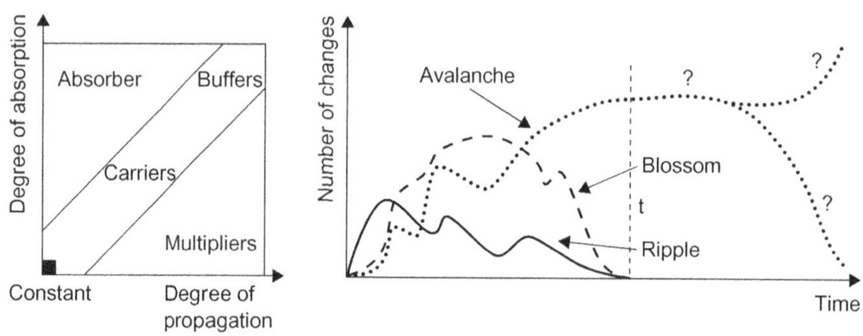

Figure 3.5
The resistance to change and the resulting patterns of process behaviour. Source: Eckert *et al.* 2004

has gone into production, many of these designers have moved on to new product. Change is a disturbance. Therefore many companies have dedicated teams to look after 'in service', i.e. in operation product. These designers have to learn about the design and understand the ramifications of the change. How well a designer can assess the effects also depends on the understanding and overview they have over the design, as illustrated in Figure 3.6. In general designers are likely to be aware of the geometric links rather than less obvious links, such as thermal links, which require a deeper understanding to recognise.

Which changes are required in the first instances comes from the user side. Users set requirements and users are also the ultimate beneficiaries of legal requirements for safety or pollution. User-generated changes often come in groups as one version of the product is modified to the next. These changes can influence each other, so that a change can be predicted in isolation but must be looked at in the context of the whole product. Change is not always a problem, but also can be seen as an opportunity to improve the product. When components are revised other features can be incorporated.

3.4 Tools for change prediction

The question is now: How can designers predict the effects of probabilistic change propagation and what tools are available to support them in this task to have better change processes as stated in section 3.2?

Computer-aided design (CAD) tools can be used for change propagation analysis. For instance, CAD solid modelling tools can identify geometrical interferences between components when a new component design is introduced. However, at each point a decision must be taken and the affected component must be redesigned.

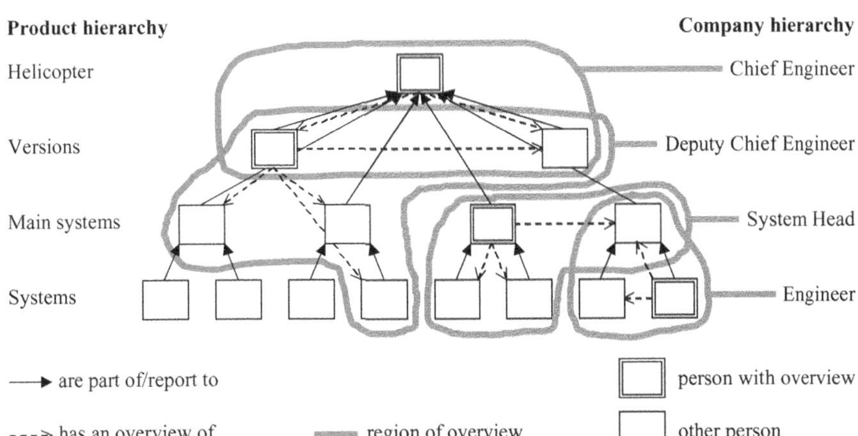

Figure 3.6 Overview in an organisation. Source: Eckert *et al.* 2004

Some academic tools are assessing the impacts of change during design by essentially working out during the design process what the consequences of potential changes later could be. Ollinger and Stahovich (2001) develop the RedesignIT tool to assess probable behavioural side-effects during a design change; Cohen *et al.* (2000) introduce a methodology called Change Favourable Representation (C-FAR) to capture possible change consequences using existing product data information.

Weber *et al.* (2003) and Conrad *et al.* (2007) describe a Property-Driven Development approach to assess the effects of changes by analysing the relationship between product behaviour and component characteristics given a set of internal dependencies and external conditions; Clarkson *et al.* (2004) develop the Change Prediction Method (CPM) which aims to identify the likelihood, impact and risk of change propagation between components. The CPM can be considered as an enhancement to the traditional Dependency Structure Matrix (DSM) analysis – see Browning (2001) for an introduction to DSM – as it examines not just the direct component linkages but also the indirect component linkages.

There are a number of attempts to model the entire spectrum of change propagation by extending the analysis to cover different concerns. For instance, Ariyo *et al.* (2004) attempt to relate the connection between Functions, Behaviours and Structure during a design change; Flanagan *et al.* (2003) introduce a mapping system to link component change propagation to their associated functions; Keller *et al.* (2007) discuss the combined used of the Contact and Channel Model (C&CM) (Matthiesen 2002) for product functional analysis and the Change Prediction Method (CPM) (Clarkson *et al.* 2004) for component change propagation analysis. More specifically, a recent attempt to consider the full effects of change propagation during concept selection has been described by Koh *et al.* (2008). This method uses the House of Quality (HoQ) (Clausing and Fey 2004) and the Change Prediction Method to break down the change propagation problem into two (see Figure 3.6).

4 The CPM tool for change prediction and visualisation

The CPM method, as described by Clarkson *et al.* (2004), is centred on the computation of indirect change propagation risks between components. The basic assumption is that, if one component changes, this change can have knock-on effects on connected components, meaning that there exists a probability that adjacent components change in response to the initiating component. These components can then, in turn, cause changes to adjacent components, so that change spreads through the system. This does not cover whole system parameters, as a model that connects all components would have no information value. Changes have been modelled as linkage types, such as mechanical, thermal or aerodynamical, to remind designers of potential connections (Jarratt 2004). The CPM method aims at identifying these 'hidden' indirect change dependencies between components and drawing the attention of the design engineers and managers responsible towards 'hidden' high-risk connections. It was shown (Jarratt 2004) that the results obtained through this method match the expectations of experienced designers and that the method was also able to predict past cases of change propagation.

Around this method, a change management methodology was developed (see Figure 3.7) which consists of three stages: building a product model, computing combined risks and analysing risks. The first stage involves the creation of a product linkage model. This model captures the components of a product and models linkages between them (Jarratt *et al.* 2004). This model is then further refined to a probabilistic model that also captures the likelihood and impact of a change propagating between each connected pair of components. These data are then used in the second stage to compute combined risk values. In the analysis stage (stage 3) these data are visualised in such a way that high-risk connections can be easily identified and acted upon.

One of the challenges of building these models to identify the right level of detail in the product breakdown is that a very detailed model is extremely difficult to handle and a too abstract model might obscure the details. During the model-building exercises, specific effort was placed on making sure that components that could be part of one or another system were clearly identified. All complex systems and thus products are inherently lattice structure, where components belong to more than one system. Here the challenge lies in disambiguating this, so that users intuitively understand the tree structure that is superimposed on the underlying lattice structure.

The CPM software tool is an implementation of the CPM methodology that supports designers in all three stages of the methodology shown in Figure 3.7. The software is a stand-alone application that is being introduced into industry. The interfaces have been designed in tight collaboration with designers in industry, and their constant feedback was used to guide the software development process.

The general strategy implemented in the software is similar to that used in explorative data-analysis tools which allow the user to explore data (i.e. the product) in an abstract way and perform most of the analyses visually using multiple linked representations. Such a strategy has proved to be successful in design (Gero and Reffat 2001) and other disciplines, such as statistics (Unwin 2000).

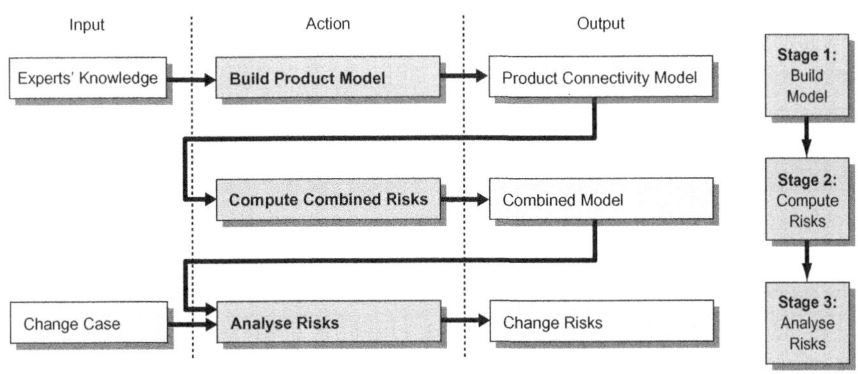

Figure 3.7 Stages of CPM change prediction method

5 Addressing complexity through the CPM tool

As we have argued in section 2.1, complexity in engineering arises from products, processes, designer, users and their interactions. These factors are closely related, and by addressing or at least expressing one of the factors others can be supported and improved. While we have developed the CPM tool to predict engineering change, the tool fundamentally maps the relationship between things and calculates how change to one element could affect another. We have also used the tool to model supply chains (Eckert *et al.* 2008) and the environmental impact of product elements.

However, in the following we shall show how a simple product model such as the CPM model can support the other sources of complexity.

5.1 Product

The CPM tool directly models the elements of the product and the connections between then. This enables designers to work on reducing the complexity within the product, by reducing the linkage between components. For example, by putting a thermal insulation coating on a component, a thermal link can be eliminated and the product can become easier to change. By identifying highly connected areas, designers can think about either merging those components into an integrated component or system or taking steps to reduce the connectivity. When a new design is created by modification the change tool allows designers to plan the changes that they want to carry out, so that can target their design effort to where they add value to the product. The change tool can also help companies in planning product platforms, where cost savings are made through an economy of scale. However, it pays companies to keep some parts flexible, which could absorb the change.

On the left of Figure 3.8(a) is a combined risk matrix for a diesel engine, which indicates the direct risk that exists between components. High risks are highlighted in red. Figure 3.8(b) shows a change propagation tree (b), which shows the many different change propagation paths that could exist, starting from a particular component, and is a representation of the product connectivity. The rings on the background show bands of risk and give a visual indication of propagated risk.

Regardless of the state of individual margins, the matrix in Figure 3.8(a) can show that some components receive a lot of change, but pass little change on, and other components do not receive a lot of change, but once they are changed cause many potential changes to other components. Figure 3.9 shows such a component classification based on change risk. This information can then be used actively to manage the change behaviour of the components. The *x*-axis shows the cumulated risk to other components resulting from a change to this component; the *y*-axis represents the cumulated risk from incoming change. Components in the top-left box have a small effect on other components but have a high risk of being changed by changes to other components (propagation absorber). Components in the bottom-right box are

(a)

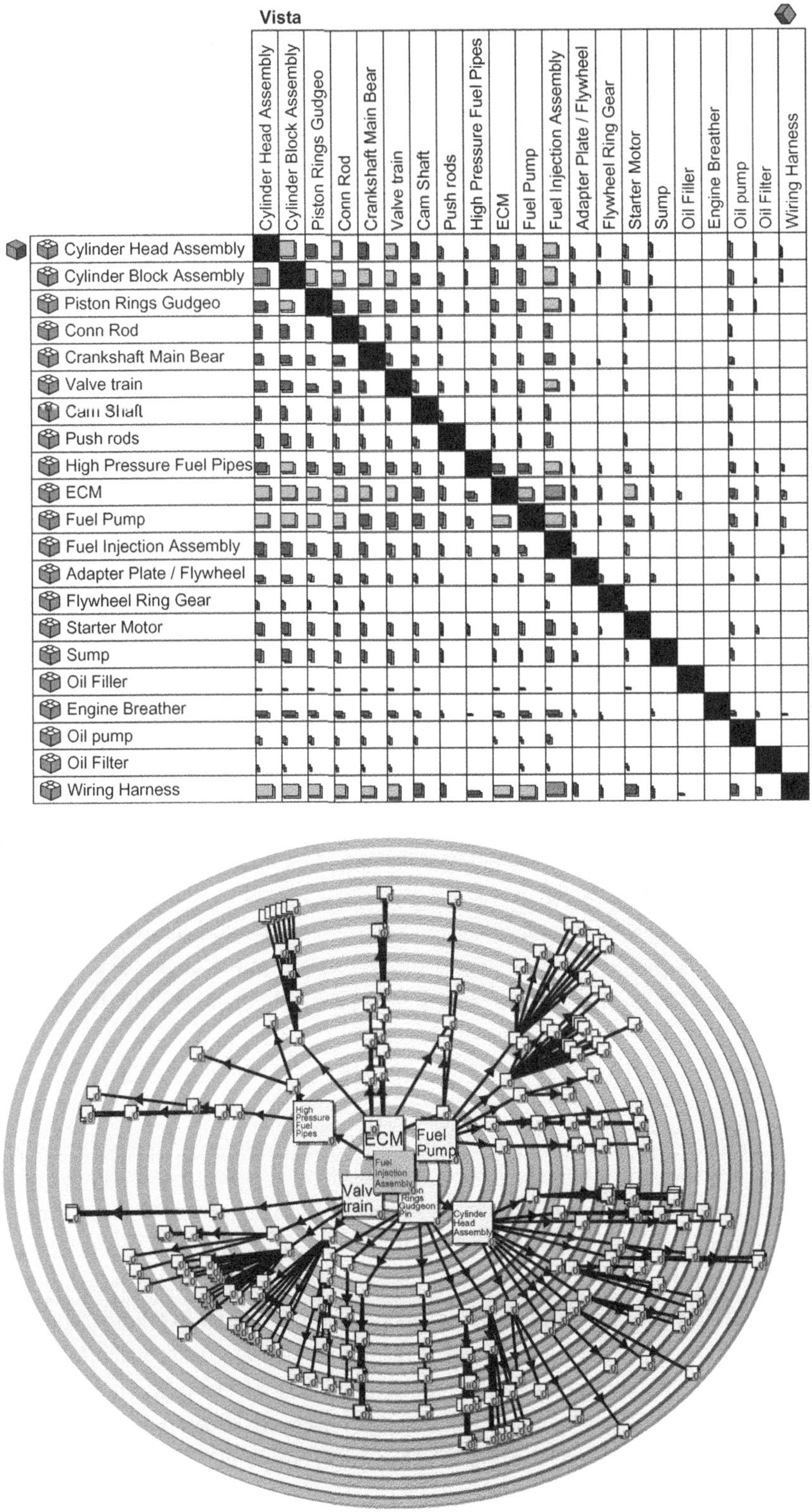

(b)

Figure 3.8
(*a*) Combined risk matrix;
(*b*) propagation tree

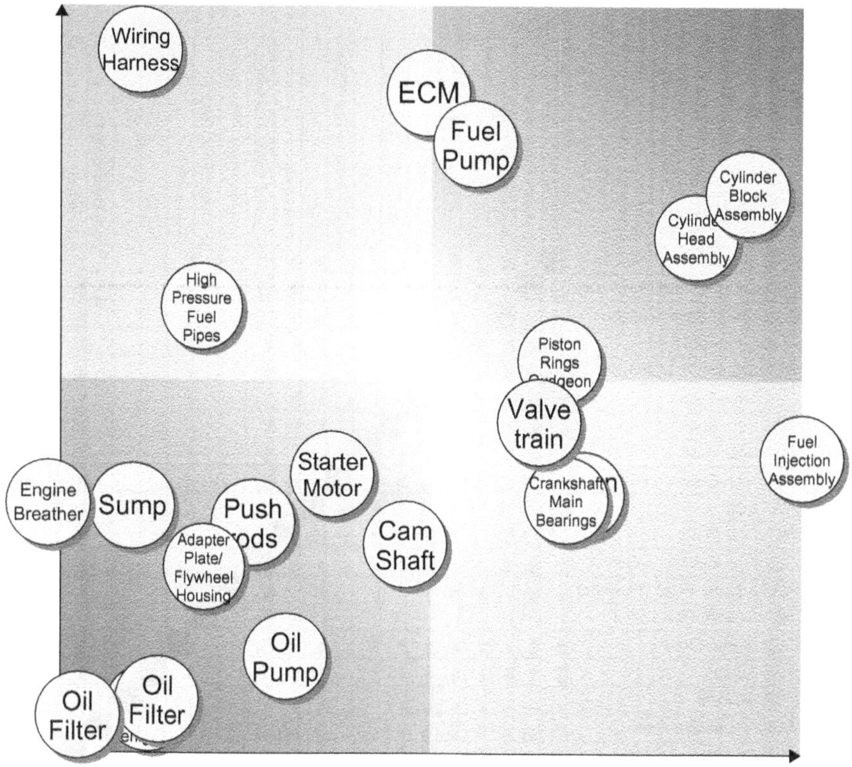

Figure 3.9 Component classification based on change risks

propagation multipliers, i.e. components that are rarely affected by other components, but changes to them require potential redesign to a number of other components. Then there are components that have a generally low impact (bottom-left box) and components that both are affected by other components and affect other components (top right). The latter components will require especially high attention from the designers, as their behaviour cannot be easily predicted.

5.2 The process

By planning the product design, it is also possible to plan the effort involved in designing a product and therefore the tasks that are associated with planning the product. Unexpected change propagation can be a major search of iteration and delays in design processes (Wynn *et al.* 2007). The risk model can play a vital role in managing this. As most companies have a joint pool of resources, delays or additional effort on one project often has knock-on effects on other projects. Late changes often require the recall of experts who have moved on to other projects. This fosters the prevalent fire-fighting culture in engineering firms, which rewards people for solving problems rather than avoiding them in the first place.

As argued by Eger *et al.* (2005), engineering freeze links products and their processes very closely. By freezing certain decisions early, it is possible to make final decision on other aspects of the product, rather than requiring iteration loops to firm up decision. However, it also means that the design of other parts can become more difficult, as the decisions that are already taken constrain the decisions that still need to be taken. Freezes are strongly influenced by the supply chain, as lead times cannot be changed. The change tools give an indication as to which freeze should be postponed, for example by incrementally releasing design information to the supplier or by keeping production in-house instead of outsourcing a component.

See Figure 3.10 for the freeze application of the CPM tool which compares change propagation results to actual design process constraints. Through the design processes, components are 'frozen', i.e. fixed either because many other decisions depend on them and the related parts can not be designed unless those decision are made, or because the component has a long lead time, i.e. takes a long time to be designed and produced. Figure 3.10 shows an analysis of the actual freeze of a particular product development process of a diesel engine and the freeze order that would arise from an analysis of the change prediction (Keller *et al.* 2008a). This assumes that those components that pass a lot of change on should be frozen early and those that receive a lot of change late. It shows that there is a high mismatch between the process (lead times) and the product (freeze order) which might lead to problems in the design process. By visualising this mismatch, designers are enabled to resolve some of the issues by, for example, reducing lead times of certain components or decreasing product connectivity to change the freeze order.

5.3 The designers

For the designers, the CPM tool is a visualisation aid that enables them to think through the changes that they want to carry out. Thinking about the consequences of change

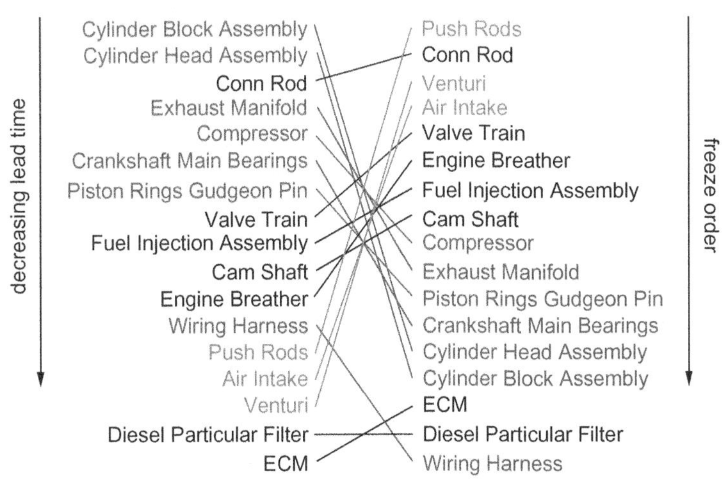

Figure 3.10 Freeze order arising from change

is essentially a large search problem. Systematically looking for all the consequences would essentially require an exhaustive search. Anecdotal evidence would indicate that designers either think through the potential consequence of a change along one chain of potential consequences – practically conducting a depth-first search – before running out of steam, or look for the places where change has occurred in the past and look for local consequence, akin to simulated annealing (Earl *et al.* 2005). In either case designers can look at all the potential consequences of a change, and therefore need guidance where they should be looking. In practice this stops companies from considering solution alternatives in detail.

The change visualisation can also act as a boundary object between (Star 1989) different groups of designers, where otherwise tacit and dispersed knowledge is made explicit and put into a notation that can be understood by several parties. In this way the tool helps not only individuals in assessing and thinking through a change, but also the interaction of teams.

5.4 The user

The user of the product is unlikely to encounter the CPM tool directly, but it can help the customer-facing part of the organisation to understand what they can offer to a customer at what cost. When the customers approach organisations, the tool can be used to cost potential customisation required to meet specific customer needs. However, the same reasoning can also be used to negotiate the needs with customers. Often customers would be just as happy with a slightly different product, which would be much cheaper. For example, one strategy that companies use is to over-engineer a product in order to standardise parts.

6 Conclusions

Engineering design is riddled with complexity and uncertainty. As man-made systems, engineering products need to meet the need of their user, but also the context of its use. Being created by people, the processes are fraught with all the richness and sub-optimality of human behaviour. Most engineering products are too complex to be designed in the head of one person, and they have to contend with all issues arising from representation as well as from team working. At the same time, some engineering designs push the limits of the technically feasibility, so that innovation-introduced risk and variability of the physical materials can cause problems on a micro-level which work up through the system.

In engineering design the goal is to reduce complexity, to find means to make problems simpler and to make solving them simpler. Design by modification is a strategy that helps designers to reduce the problem they need to solve by re-using parts of existing solutions, but it also ties different design problems together. The CPM tool illustrates how engineering designers approach the tool and methods as a means to reduce this complexity.

References

Alligood, K. T., Sauer, T. and Yorke, J. A. (2001), *Chaos: An Introduction to Dynamical Systems*, London: Springer Verlag.

Ariyo, O. O., Eckert, C. M. and Clarkson, P. J. (2004), 'Tolerance margins as constraining factors of changes in complex products' 5th Integrated Product Development Workshop (IPD 2004), Schonebeck/Bad Salzelmen bei Magdeburg, Germany: CD-ROM.

Brown, J. (2006), *Managing Product Relationships: Enabling Iteration and Innovation in Design*, Boston, Mass.: AberdeenGroup.

Browning, T. R. (2001), 'Applying the design structure matrix to system decomposition and integration problems: a review and new directions', *IEEE Transactions on Engineering Management*, 48 (3): 292–306.

Bucciarelli, L. L. (1996), *Designing Engineers*, Cambridge, Mass.: MIT Press.

Clark, K. B. and Fujimoto, T. (1991), *Product Development Performance: Strategy, Organization and Management in the World Auto Industry*, Boston, Mass.: Harvard Business School Press.

Clarkson, P. J., Simons, C. and Eckert, C. M. (2004), 'Predicting change propagation in complex design'. *Journal of Mechanical Design*, 126 (5): 765–97.

Clausing, D. and Fey, V. (2004), *Effective Innovation: The Development of Successful Engineering Technologies*, New York: John Wiley.

Cohen, T., Navathe, S. B. and Fulton, R. E. (2000), 'C-FAR, change favourable representation', *Computer-Aided Design*, 32 (5): 321–38.

Conrad, J., Deubel, T., Kohler, C., Wanke, S. and Weber, C. (2007), 'Change impact and risk analysis (CIRA): combining the CPM/PDD theory and FMEA methodology for an improved engineering change management', International Conference on Engineering Design (ICED'07), Paris: CD-ROM.

Cooper, K. G. (1993), 'The rework cycle: benchmarks for the project manager', *Project Management Journal*, 24 (1): 17–21.

de Weck, O. L. and Suh, E. S. (2006), 'Flexible product platforms: framework and case study', ASME 2006 Design Engineering Technical Conferences, Philadelphia, Pa: CD-ROM.

Deubzer, F., Kreimeyer, M., Rock, B. and Junior, T. (2005), 'Der Änderungsmanagement Report 2005', *CiDaD Working Paper Series*, 1 (1): 2–12.

Earl, C., Eckert, C. M. and Clarkson, P. J. (2005), 'Predictability of change in engineering: a complexity view', ASME 2005 Design Engineering Technical Conferences, Long Beach, Calif., USA: CD-ROM.

Earl, C., Johnson, J. and Eckert, C. M. (2005), 'Complexity', in P. J. Clarkson and C. M. Eckert (eds), *Design Process Improvement: A Review of Current Practice*, London: Springer Verlag.

Eckert, C. M., Clarkson, P. J. and Zanker, W. (2004), 'Change and customisation in complex engineering domains'. *Research in Engineering Design*, 15 (1): 1–21.

Eckert, C. M., Jowers, I. and Clarkson, P. J. (2007), 'Knowledge requirements over long product lifecycles', International Conference on Engineering Design (ICED 07), Paris: CD-ROM.

Eckert, C. M., Zolghadri, M., Keller, R. and Clarkson, P. J. (2008), 'Indirect connections in a supply chain: visualisation and analysis', *10th International Design Structure Matrix Conference, DSM'08*, Stockholm: 95–104.

Eger, T., Eckert, C. M. and Clarkson, P. J. (2005), 'The role of design freeze in product development', International Conference on Engineering Design (ICED 05), Melbourne, Australia: CD-ROM.

Flanagan, T., Eckert, C. M., Eger, T., Smith, J. and Clarkson, P. J. (2003), 'A functional analysis of change propagation', *International Conference on Engineering Design (ICED 03)*, Stockholm: 441–2.

Flanagan, T., Eckert, C. M., Keller, R. and Clarkson, P. J. (2006), 'Bridging the gaps between project plans and reality: the role of overview' *6th International Symposium on Tools and Methods of Competitive Engineering (TMCE 2006)*, Ljubljana: 105–16.

Fricke, E., Gebhard, B., Negele, H. and Igenbergs, F. (2000), 'Coping with changes: causes, findings, and strategies', *Systems Engineering*, 3: 169–79.

Gero, J. S. and Reffat, R. M. (2001), 'Multiple representations as a platform for situated learning systems in designing', *Knowledge-Based Systems*, 14: 337–51.

Harrison, A. (2006), 'Design for service: harmonising product design with a service strategy', GT2006 ASME Turbo Expo 2006: Power for Land, Sea and Air, Barcelona.

Hirsch, C. (2007), *Numerical Computation of Internal and External Flows: Fundamentals of Computational Fluid Dynamics*, Oxford: Butterworth-Heinemann.

Jarratt, T. (2004), 'A model-based approach to support the management of engineering change', PhD dissertation, University of Cambridge.

Jarratt, T., Eckert, C. M. and Clarkson, P. J. (2004), 'Development of a product model to support engineering change management', *Tools and Methods of Competitive Engineering (TCME 2004)*, Lausanne: 331–42.

Jarratt, T., Eckert, C. M., Weeks, R. and Clarkson, P. J. (2003), 'Environmental legislation as a driver of design', International Conference on Engineering Design (ICED 03), Stockholm: CD-ROM.

Julier, G. (2000), *The Culture of Design*, London: Sage.

Kauffman, S. and Macready, W. (1995), 'Technological evolution and adaptive organizations, *Complexity*, 1 (2): 26–43.

Keller, R., Alink, T., Pfeifer, C., Eckert, C. M., Clarkson, P. J. and Albers, A. (2007), 'Product models in design: a combined use of two models to assess change risks', International Conference on Engineering Design (ICED'07), Paris: CD-ROM.

Keller, R., Eckert, C. M. and Clarkson, P. J. (2008a), 'Determining component freeze order: a redesign cost perspective using simulated annealing', ASME 2008 International Design Engineering Technical Conferences, New York: CD-ROM.

Keller, R., Eckert, C. M. and Clarkson, P. J. (2008b), 'Through-life change prediction and management', International Conference on Product Lifecycle Management, Seoul: CD-ROM.

Koh, E. C. Y., Keller, R., Eckert, C. M. and Clarkson, P. J. (2008), 'Influence of feature change propagation on product attributes in concept selection', Design 2008, Dubrovnik.

McMahon, C. A. (1994), 'Observations on modes of incremental change in design', *Journal of Engineering Design*, 5 (3): 195–209.

Matthiesen, S. (2002), 'A contribution to the basis of the element model "Working Surface Pairs and Channel and Support Structures" on the correlation between layout and function of technical systems', PhD dissertation. Technical University Karlsruhe.

Maull, R., Hughes, D. and Bennett, J. (1992), 'The role of the bill-of-materials as a CAD/CAPM interface and the key importance of engineering change control', *Computing and Control Engineering Journal*, 3 (2): 63–70.

Mont, O. K. (2002), 'Clarifying the concept of product-service system', *Journal of Cleaner Production*, 10 (3): 237–45.

Ollinger, G. A. and Stahovich, T. F. (2001), 'RedesignIt – a constraint-based tool for managing design changes', ASME 2001 Design Engineering Technical Conferences, Pittsburgh, Pa: CD-ROM.

Otto, K. and Wood, K. (2001), *Product Design: Techniques in Reverse Engineering, Systematic Design, and New Product Development*, New York: Prentice-Hall.

Pahl, G. and Beitz, W. (1996), *Engineering Design: A Systematic Approach*, London: Springer-Verlag.

Simon, H. A. (1998), *The Sciences of the Artificial*, Cambridge, Mass.: MIT Press.

Simpson, T. W., Marion, T., de Weck, O. L., Hollta-Otto, K., Kokkolaras, M. and Shooter, S. (2006), 'Platform-based design and development: current trends and needs in industry', ASME 2006 Design Engineering Technical Conferences, Philadelphia, Pa: CD-ROM.

Star, S. L. (1989), 'The structure of ill-structured solutions: boundary objects and heterogeneous distributed problem solving', in L. Gasser and M. N. Huhns (eds), *Distributed Artificial Intelligence*, London: Pitman, 37–54.

Suh, N. P. (2001), *Axiomatic Design: Advances and Applications*, New York: Oxford University Press.

Ulrich, K. T. and Eppinger, S. D. (2004), *Product Design and Development*, New York: McGraw-Hill.

Unwin, A. R. (2000), 'Using your eyes: making statistics more visible with computers', *Computational Statistics and Data Analysis*, 32: 303–12.

Vincenti, W. (1990), *What Engineers Know and How They Know It*, Baltimore, Md: Johns Hopkins University Press.

Weber, C., Werner, H. and Deubel, T. (2003), 'A different view on product data management/product life-cycle management and its future potentials', *Journal of Engineering Design*, 14 (4): 447–64.

Wynn, D. C., Eckert, C. M. and Clarkson, P. J. (2007), 'Modelling iteration in engineering design', *16th International Conference on Engineering Design (ICED'07)*, Paris: 693–4.

This page intentionally left blank

4

Using complexity science framework and multi-agent technology in design

George Rzevski

1 Introduction

Traditionally the aim of designers has been to design organisations, processes or products that perform according to a given specification. Organisations, such as businesses, are expected to work in accordance with statutes, rules and regulations, and every employee is given a detailed job specification. Computer programs are designed to follow precise instructions written by programmers without any deviation. Dynamic systems, such as turbo-machinery, are expected to be stable, which means when disturbed to return to equilibrium.

This may have been useful under stable market conditions of the industrial age when there was an expectation that a business would function in the same way for a substantial period of time and when demand for a particular product – say, a car – would not substantially change for at least five years. And it is still valid for static products like bridges and tunnels, which are expected to serve for very long periods of time.

There is, however, a class of systems for which traditional design methods are of dubious value. Organisations, processes and products that are subject to frequent changes of demand should be designed to be *adaptable* to changes in their environments rather than stable. Consider also systems that are subject to threats, e.g., threats of viruses to software, or threats of terrorist attacks to critical installations such as nuclear power stations. Should we not design into these systems an 'immune system' that will dynamically regroup and redeploy internal resources to fight attackers? Do we know how to design vulnerable systems to be *resilient* to attacks?

My thesis is that only *complex* organisations, processes and products can be adaptive and resilient. The key skill we should learn therefore is the skill of designing *complexity* into organisations, processes and products.

Let us first refresh our understanding of complexity by looking at it from the point of view of a designer.

2 What is complexity?

The following three paragraphs from Wikipedia (http://en.wikipedia.org/wiki/Complexity) are a good introduction to the concept of complexity.

> Complexity has always been a part of our environment, and therefore many scientific fields have dealt with complex systems and phenomena. Indeed, some would say that only what is somehow complex – what displays variation without being random – is worthy of interest.
>
> The use of the term complex is often confused with the term complicated. To understand the differences, it is best to examine the roots of the two words. 'Complicated' uses the Latin ending 'plic' that means, 'to fold' while 'complex' uses the 'plex' that means, 'to weave'. Thus a complicated structure is one that is folded with hidden facets and stuffed into a smaller space. On the other hand, a complex structure uses interwoven components that introduce mutual dependencies and produce more than a sum of the parts . . . This means that complex is the opposite of independent, while complicated is the opposite of simple.
>
> While this has led some fields to come up with specific definitions of complexity, there is a more recent movement to regroup observations from different fields to study complexity in itself, whether it appears in anthills, human brains, or stock markets.

Following the train of thought suggested above, the intuitive interpretation of the term *complex* as 'difficult to understand' is correct as long as we accept that the main reason for the difficulty is the interdependence of constituent components.

An example that immediately comes to mind is the Internet-based global market, where consumers and suppliers are trading, each pursuing their own goals and targets, and where the overall distribution of resources to demands emerges from individual transactions rather than according to a given plan.

On a smaller scale we have a complex situation associated with availability of aircraft where there is a need to mange interaction between aircraft failures, aircraft locations, maintenance and repair personnel schedules, and supply of spare parts.

According to Prigogine (1997), a system is complex if its global behaviour *emerges* from the interaction of local behaviours of its components (the system creates a new order). Prigogine in his writings emphasises that the behaviour of a complex system cannot be predicted and that, in general, the future is not given (Prigogine 2003); it is being created by actions of all those that participate in the working of the Universe. He discusses examples of complex systems from physics and chemistry, including molecules of air subjected to a heat input, autocatalytic chemical processes and self-reproduction of cells. Emergent behaviour of complex systems is widely covered in literature (Holland 1998) and applied to many domains, including economics (Beinhocker 2006).

To locate complex systems on a map of predictability, I proposed (Rzevski 2008) the following system classification (Table 4.1), in which complex systems are placed between random and stable systems.

3 The key elements of complexity

Let us carefully examine the key elements of complexity, emphasising those that are essential for the design of artificial complex systems (Rzevski and Skobelev 2007).

1. Perhaps the most important feature of complex systems is that *decision-making is distributed* rather than centralised. Complex systems consist of interconnected autonomous decision-making elements, often called *agents*, capable of communicating with each other. There is no evidence of centralised control.
2. The autonomy of agents is not total. Every complex system has some global and/or local principles, rules, laws, or algorithms for agents to follow. The important point to remember is that agents' behaviour is never completely defined by these rules – they always have alternative possible local behaviours. In other words, complex systems always have a *variety* of possible behaviours and *uncertainty* which behaviour will be executed. The degree of freedom that is given to agents (decision-makers) determines the system's ability to self-organise and evolve. When uncertainty is insignificant, the system behaves predictably and lacks capabilities for self-organisation. When uncertainty is equal to 1, the

Table 4.1 A classification of systems

CLASSES/ FEATURES	RANDOM SYSTEMS	COMPLEX SYSTEMS	STABLE SYSTEMS	ALGORITHMIC SYSTEMS
Predictability	Total uncertainty	Considerable uncertainty	No uncertainty	No uncertainty
Behaviour	Random	Emergent	Planned	Deterministic
Norms of behaviour	Total freedom of behaviour	Some external guidance is essential	Governed by laws and regulations	Follows instructions
Degree of organisation	None	Self-organisation	Organised	Rigidly structured
Degree of control	None	Self-control by self-organisation	Centralised control	No need for control
Irreversible changes	Random changes	Co-evolves with environment	Small temporary deviations possible	None
Operating point	None	Operates far from equilibrium	Operates at an equilibrium	Operates according to the specification

system is chaotic (random). The adaptive complex systems operate 'at the edge of chaos' or 'far from equilibrium'. The occurrence of events that affect their behaviour is so frequent that there is no time for the system to return to its equilibrium.

3. Global behaviour of a complex system *emerges* from the interaction of constituent agents. However, because the decision-making freedom of agents is restricted, complex systems exhibit *patterns* of behaviour. Designers have a choice here. The degree of uncertainty can be adjusted to force the system to follow specified broad patterns. The complete predictability should not be aimed for – it would prevent the system from self-organising and adapting.
4. Complex systems are non-linear: the smallest external effects may cause large-scale shifts in system behaviour, the phenomenon known as *butterfly effect* (e.g. as in climate systems) or as *self-acceleration* (e.g. as in chain reaction in atomic explosions). Also, complex systems exhibit *autocatalytic* behaviour, that is the ability to create new structures without any external help (e.g. creation of organic structures from non-organic materials, under certain thermal conditions).
5. The distribution of decision-making implies interconnectedness of decision-making elements (agents). The links between agents can be strong or weak or nonexistent. The type of link between agents determines the responsiveness of the system when disturbed. Designers can weaken certain links between agents to reduce time required for ripples caused by a chain of changes to settle down.
6. The autonomy implies intelligence. Intelligence implies knowledge and a capability of applying knowledge to resolve uncertainty.

4 Technology for designing complexity into organisations, processes and products

The most effective technology for constructing complex systems which exhibit all features described in the previous section is multi-agent software (Rzevski *et al.* 2007).

In contrast to conventional software such as centralised schedulers, planners and optimisers, which from the start to the end follow algorithms, multi-agent software works primarily by exchanging messages: Intelligent Software Agents negotiate deals with each other, always consulting problem domain knowledge assembled in Ontology. Negotiations are conducted by a concurrent and asynchronous exchange of messages. The system is event-driven: it rapidly self-organises to accommodate events that affect its operation.

Problem domain knowledge is elicited and represented as a semantic network in which concepts (classes of objects) are nodes and relations between concepts are links. Each object is characterised by attributes and rules guiding its behaviour. Such a conceptual knowledge repository is called *Ontology*.

A real-life problem situation is represented as a virtual network of instances of objects defined in Ontology and their relations. Such a problem description is called a *Scene*.

The elementary computational element is called an *Agent*. An agent is a computer program capable of solving the problem at hand by consulting Ontology and using knowledge thus acquired to negotiate with other agents how to change the current Scene and turn it from the description of the problem into the description of a solution. Agents solve problems in cooperation and/or competition with other agents. As *Events* (new orders, failures, delays) affecting the problem domain occur, agents amend the current scene to accommodate the event, thus achieving *Adaptability*.

An agent is assigned to each object participating in the problem-solving process (and represented in the scene) with a task of negotiating for its client (object) the best possible service conditions. For example, Passenger Agents and Seat Agents will negotiate takeoff/landing times and seat prices for requested flights. Closing a deal between a Passenger Agent and a Seat Agent indicates that a full, or at least partial, matching between *Demand* and an available *Resource* has been achieved. In case of a partial matching (e.g. a passenger agrees to accept a later takeoff time but it is not pleased), his Agents may attempt to improve the deal if a new opportunity presents itself at a later stage (e.g. if other passengers on the same flight agree an earlier take-off time). The process continues as long as it is necessary to obtain full matches, or until the occurrence of the next event (say, a new request for a seat) which requires agents to reconsider previously agreed deals.

Agent negotiations are informed by domain knowledge from Ontology, which is far more comprehensive than 'rules' found in conventional schedulers and normally includes expertise of practising operators. Not all of this knowledge is rigid: certain constraints and if-then-else rules may be considered as recommendations and not as instructions, and agents may be allowed to evaluate their effectiveness and decide if they should be used. In some cases, agents send messages to users asking for approval to ignore ineffective rules or to stress non-essential constraints.

The power of agent-based problem-solving is particularly evident when the problem contains a very large number of objects with a variety of different attributes; when there is a frequent occurrence of unpredictable events that affect the problem-solving process; and when criteria for matching demands to resources are complex (e.g. balancing risk, profits and level of services, which may differ for different participants).

As the process is incremental, a change of state of one agent may lead to changes of states of many other agents. As a result, at some unpredictable moment in time a spontaneous self-accelerated chain reaction of state changes may take place, and after a relatively short transient time the overall structure will switch its state practically completely. Once the resulting structure has settled, the incremental changes will continue.

Agent-based problem-solving process appears to exhibit autonomy and intelligence, known as 'emergent intelligence' and is therefore often called 'intelligent problem-solving process'.

4.1 Architecture

A multi-agent software comprises the following key components: (*a*) Multi-Agent Engine, which provides runtime support for agents; (*b*) Virtual World, which is an environment in which agents cooperate and compete with each other as they construct and modify the current scene; (*c*) Ontology, which contains conceptual problem domain knowledge network; and (*d*) Interfaces.

4.2 How multi-agent software works

Software consists of a set of continuously functioning agents that may have contradictory or complimentary interests. Basic roles of agents, based on extended Contract Net protocol, are Demand and Supply roles: each agent is engaged in selling to other agents its services or buying services it needs (Passenger Agents buy seats and Seat Agents sell them).

Current problem solution (current scene) is represented as a set of relations between agents which describe the current matching of services; for example, a schedule is a network of passengers, seats, aircraft and flights, and relations between them.

The arrival of a new event in the system is triggered by the occurrence of a change in the external world; for example, when a passenger requests a seat on a particular flight, a Seat Request Event is triggered in the system.

The agent representing the object affected by the new event undertakes to find all affected agents and notify them of the event and the consequences (e.g. the agent of the failed aircraft undertakes to find Passenger Agents linked to the failed flight and inform them that the flight is not available; the Aircraft Agent breaks the relevant relations and frees the Passenger Agents to look for other available flights).

The process of problem-solving can run in parallel and asynchronously and, as a consequence, simultaneously by several active participants; for example, passengers that arrived at the website to book a flight simultaneously can all immediately start searching for suitable seats. All aircraft assigned to flights can start immediately looking for free pilots. This feature is very effective because it eliminates a laborious building of flight schedules only to find out that pilots are not available for all selected flights.

The driving force in decision-making is often the presence of conflicts, which have to be exposed and settled by reconsidering previously agreed matches; for example, if a new flight finds out that the takeoff time-slot it needs is already occupied, negotiations on conflict resolution start and, as a result, previously agreed flight-slot matches are adjusted (the takeoff time-slot is moved to accommodate both flights) or broken (the time-slot is freed). This capability to make local adjustments before introducing big changes is what makes agent-based problem-solving so much more powerful in comparison with object-oriented or procedure-based methods.

A multi-agent system is in a perpetual state of processing – either reacting to the arrival of new events or improving the quality of previously agreed matches.

The stable solution, when there are no agents that can improve their states and there are no new events, is hardly ever reached (agents are perpetually operating 'far from equilibrium').

Solutions developed using multi-agent software fall into the class of open, non-linear and dissipative systems. As the number of relations increases in the system, the level of complexity of the resulting network goes up and, at a certain point, the need may arise to appoint additional agents to represent certain self-contained parts of the network whose nodes are already represented by agents. The increased complexity of solution structures may result in the creation of loops, and the system may find itself in a local optimum. To avoid being stuck in a local optimum, agents are from time to time given power to seek alternative solutions proactively. Attempts to avoid local minima are random (mutations).

Multi-agent systems can learn from experience as follows. Logs of agent negotiations are analysed with a view to discovering patterns linking individual agent decisions and successes/failures of the agent negotiation process. In future negotiations, patterns leading to failures are avoided.

The pattern-discovery process is itself agent-based. An agent is assigned to each data element with a task of searching for similar data elements to form clusters. An agent is assigned to each new cluster with a task of attracting data elements that meet cluster membership criteria. Finally, clusters are represented as 'if-then-else rules'.

5 Designing for adaptability

During the last ten years I have accumulated a rich practical experience in designing complexity into organisations, business processes and products with a view to making them adaptable. Based on this experience, I propose here a set of key fundamental principles of the design for adaptability.

5.1 Key design principles

1. To design an organisation or a (business) process to be adaptive, the most important move is to make decision-making distributed. Centralised control and command system should be replaced by teamwork.
2. Team decision-making should be guided by company policies, rules and regulations rather than directed and controlled by a chief executive.
3. Restrictions to freedom of decision-making of agents should be tuned finely by adjusting the quantity and rigidity/flexibility of policies, rules and regulations.
4. Provision should be made for autocatalytic properties and for self-acceleration of processes that lead to effective adaptation. This can be achieved by allowing a certain amount of experimentation with bold, innovative ideas and, most importantly, by planning a variety of alternative ways of achieving the specified goal.

5. Strengths of links between teams and between members of each team should be carefully planned to enable rapid re-adjustment in reaction to important events without excessive negotiation ripples. Typically the communication links between members of a team must be strong while the strength of links between teams must be graded to reflect the closeness of team objectives.
6. Enterprise ontology should be designed to contain domain knowledge required for effective operation of the organisation.
7. All human decision-makers should be supported by highly adaptive agent-based systems and all routine decisions delegated to intelligent multi-agent technology connected to common Enterprise Network.
8. Physical objects (aircraft, cars, trucks, ships, cargoes, pallets, parcels, components, etc.) constituting parts of the organisation should be supplied with RFID tags and their environments (hangars, workshops, plants, warehouses, cargo holds) equipped with RFID readers so that physical objects can be connected to the same Enterprise Network.

Principles for designing adaptive products are exactly the same if we accept that words used above, such as 'teamwork', 'chief executive', 'organisation', are used as metaphors meaning 'collaboration of components', 'centralised controller' and 'product'.

5.2 Organisational and process design cases

Systems that have been designed under my supervision, or with my involvement, using the above principles are described in some detail in Rzevski *et al.* (2006), Andreev *et al.* (2007), Rzevski *et al.* (2003), Minakov *et al.* (2007) and Rzevski *et al.* (2007). Advantages of adaptability in comparison with rigid systems, such as ERP (Enterprise Resource Planning), are described in a popular format in Brace and Rzevski (1998).

The list is substantial and includes realtime, adaptive multi-agent systems for: managing 10 per cent of world tanker capacity for global crude oil transportation (in use); managing 2,000 taxis and other service vehicles in London (in use); managing an extensive road logistic system across the United Kingdom (in use); managing social entitlements of citizens in a very large region (in use); managing distribution of rental cars across Europe for a major global car rental organisation (successful trials; in the commissioning stage); managing a car manufacturing system (prototype); simulating virtual enterprises (prototype); managing document flow for a major insurance company (prototype); managing all business processes of a new aviation company (in the design stage); managing a catering supply chain (in the design stage).

To illustrate the power of agent-based adaptive systems, let me outline the complexity of the design problem I am handling at present. The goal is to design an adaptive organisation based on teamwork, and a supporting intelligent multi-agent management system, which will make autonomously all operational decisions (how much to charge a customer for a flight, which pilot, aircraft, ground staff will be assigned to which duty, etc.) and manage domain knowledge required for strategic decisions (on expansion, on market penetration, on increasing business value) for a brand-new enterprise. The enterprise network will enable rapid interaction of twelve multi-agent modules, including simulators, several schedulers, a demand-forecasting system, a human resource management system, and customer relations management system, and will maintain integrity of all enterprise data and several enterprise ontology. The system is being designed to handle 4,000 travel requests a day, to book 400 taxi seats/flights a day, and to schedule or reschedule a large fleet of small aircraft, flights, crews, ground staff, aircraft maintenance, fuel supply, etc., every seven seconds, which is a task that would be impossible to achieve without agent-based technology.

5.3 Product design cases

I am not aware of any fully adaptive products in operation. However, I have researched, simulated or prototyped a number of distributed, adaptive products following the design principles outlined above, including: a machine tool; an intelligent geometry compressor; an autonomous parcel distribution system; and an intelligent family of robots (Rzevski 1998, 2003).

Perhaps the boldest idea is to design a compressor with moving individual vanes capable of autonomously and dynamically positioning themselves at the optimum angle whenever the operating point of the compressor changes. A software agent is assigned to each individual moving vane equipped with a pressure sensor. As pressure on the vane changes, the Vane Agent negotiates with agents of other vanes how to change vane angles to achieve the optimum pressure distribution along the stator. Before making a decision, agents consult domain knowledge stored in individual agents' minds, which can be updated without interrupting work of agents, to fine-tune compressor operation. A very successful simulation (Morgan *et al.* 2004) showed that the compressor with autonomous vanes, when coupled with an aircraft jet engine, is fully adaptive to sudden changes of loads and is able to prevent stalling of jets caused by lack of air intake.

Replacing a robot by a family of smaller robots illustrates the advantages of designing complexity into artefacts even better. To avoid disasters that ruined both American and British Mars exploration robots (the first died from the accumulation of space dust on its solar cells after a week in space, and the latter fell into a crevice on landing and was immediately lost), I proposed to design a family of five smaller robots capable of cleaning each other, rescuing members of the family from disasters and, more importantly, able to complete their task successfully even if one or two of the family members were disabled.

A family of robots is an adaptive distributed system which incorporates all critical complexity features listed earlier in this paper. All decisions are executed after a process of consultation and negotiation among members of the family. There is no 'senior' robot ordering others what to do. Each robot is controlled by a set of interacting swarms of agents. Agents consult domain knowledge before making decisions. A copy of domain knowledge is stored in each robot's ontology, making it capable of undertaking any task within domain boundaries. The family represents a 'swarm of interacting swarms' of agents and therefore exhibits a considerable emergent intelligence. Robots are trained to help each other, share the workload and self-organise the team if a member is disabled without losing ability to achieve the goal.

6 Conclusions

A set of principles for designing realtime, adaptive organisations, processes and products is articulated in this paper, based on experimenting with very-large-scale multi-agent software systems. These principles appear to be counter-intuitive – they support the creation of complex adaptive systems instead of, currently favoured, rigid predictable structures. The acceptance of this approach requires a complete change of the designer's mindset, which can be achieved only by extensive re-education. Principles have been tested in commercial projects and found very effective in domains characterised by frequent changes of operating conditions.

References

Andreev, M., Rzevski, G., Skobelev, P., Shveykin, P., Tsarev, A. and Tugashev, A. (2007), 'Adaptive planning for supply chain networks', In V. Marik, V. Vyatkin and A. W. Colombo (eds), *Holonic and Multi-Agent Systems for Manufacturing*, Third International Conference on Industrial Applications of Holonic and Multi-Agent Systems, HoloMAS 2007, Regensburg, September 2007, pp. 215–25. Springer LNAI 4659.

Beinhocker, E. (2006), *The Origin of Wealth: Evolution, Complexity and the Radical Remaking of Economics*, Cambridge, Mass.: Harvard Business School Press.

Brace, G. and Rzevski, G. (1998), 'ERP – Elephants Rarely Pirouette', *Logistics Focus*, 6 (9).

Holland, J. (1998), *Emergence: From Chaos to Order*, Oxford: Oxford University Press.

Minakov, I., Rzevski, G., Skobelev, P. and Volman, S. (2007), 'Creating contract templates for car insurance using multi-agent based text understanding and clustering', In V. Marik, V. Vyatkin and A. W. Colombo (eds), *Holonic and Multi-Agent Systems for Manufacturing*, Third International Conference on Industrial Applications of Holonic and Multi-Agent Systems, HoloMAS 2007, Regensburg, September 2007, pp. 361–71. Springer LNAI 4659.

Morgan, G., Rzevski, G. and Wiese, P. (2004), 'Multi-agent control of variable geometry axial turbo compressors', *Journal of Systems and Control Engineering*, 13 (218): 157–71.

Prigogine, I. (1997), *The End of Certainty: Time, Chaos and the New Laws of Nature*, New York: Free Press.

Prigogine, I. (2003), *Is Future Given?*, Hackensack, NJ: World Scientific Publishing Company.

Rzevski, G. (1998), 'Engineering Design for the Next Millennium: The Challenge of Artificial Intelligence', the 86th Thomas Hawksley Memorial Lecture, IMechE, 9 December.

Rzevski, G. (2003), 'On conceptual design of intelligent mechatronic systems', *Mechatronics*, 13: 1029–44.

Rzevski, G. (2008), 'Investigating current social, economic and educational issues using framework and tools of complexity science', *Journal of the World University Forum*, 1 (2).

Rzevski, G., Himoff, J. and Skobelev, P. (2006), 'Magenta technology: a family of multi-agent intelligent schedulers', Workshop on Software Agents in Information Systems and Industrial Applications (SAISIA). February. Fraunhofer IITB.

Rzevski, G. and Skobelev, P. (2007), 'Emergent intelligence in large scale multi-agent systems', *International Journal of Education and Information Technology*, 1 (2): 64–71.

Rzevski, G., Skobelev, P. and Andreev, V. (2007), 'MagentaToolkit: a set of multi-agent tools for developing adaptive real-time applications', In V. Marik, V. Vyatkin and A. W. Colombo (eds), *Holonic and Multi-Agent Systems for Manufacturing*, Third International Conference on Industrial Applications of Holonic and Multi-Agent Systems, HoloMAS 2007, Regensburg, September, pp. 303–14. Springer LNAI 4659.

Rzevski, G., Skobelev, P., Batishchev, S. and Orlov, A. (2003), 'A framework for multi-agent modelling of virtual organisations', In L. M. Camarinha-Matos and H. Afsarmanesh (eds), *Processes and foundations for Virtual Organisations*, Dordrecht: Kluwer, pp. 253–60.

Rzevski, G., Skobelev, P., Minakov, I. and Volman, S. (2007), 'Dynamic pattern discovery using multi-agent technology', *Proceedings of the 6th WSEAS International Conference on Telecommunications and Informatics* (TELE_INFO '07), Dallas, Tex., 22–4 March, pp. 75–81.

This page intentionally left blank

5 Complexity and coordination in collaborative design

Katerina Alexiou

1 Introduction

Design processes are complex in nature and commonly involve multiple participants or agents who bear individual (and often conflicting) views, goals, expertise, knowledge and models. Constructing design solutions necessitates the coordination of distributed knowledge, processes and decisions. This chapter suggests that coordination is a suitable concept for understanding design as a multi-agent process, and also for capturing its generative and creative character. In this chapter, the theoretical understanding of design as coordination is informed by computational modelling and simulation. In particular, simulation is used as a means to think through the dimensions and characteristics of coordination, and construct a coherent framework for its understanding.

2 Multi-agent design: collaboration and coordination

Design is usually defined as a purposeful human activity (Simon 1969; Rosenman and Gero 1998) which moves from some perceived need for change towards a new (previously unknown) state that satisfies this need. But goal-seeking and goal articulation relate to problem-*making*, which is an integral part of design: according to Rittel and Webber (1984: 137), 'problem understanding and problem resolution are concomitant to each other'. Design problems are often understood and characterised as 'wicked' problems, and one of the main reasons is that design knowledge and decision-making are '*distributed* among many people, in particular among those who are likely to become affected by the solution' (Rittel and Webber 1984: 320 – my emphasis). The notion that design problems and design solutions are formed together or co-evolve during the design process has been extensively discussed as a critical aspect of creative design (Maher 1994; Smithers 1998; Dorst and Cross 2001). However, the issue of co-evolution has mainly been approached from the perspective of the individual designer or agent (even in the cases where design is taken as a process that unravels within a social context), and so a better understanding of this notion in collaborative design is needed.

2.1 Collaboration in design

Domeshek *et al.* notice: 'The lone design genius, if not mythical or completely extinct, is surely on the endangered species list. Nowadays, significant design projects require teams of designers coordinating their varied expertise to arrive at effective design solutions' (Domeshek *et al.* 1994: 143). The realisation that group processes are vital in most design projects has led to an increasing interest in collaborative design. Although there are different approaches to collaborative design, some common dimensions can be identified.

The first important dimension is that *knowledge is distributed* between the members of the design group. Kvan (1999: 62) argues: 'The knowledge required to produce an architectural design lies beyond the realm of one individual. Deriving a solution to an architectural design problem draws upon the collective knowledge of the group, the team, gathered for the project.' Note that this 'social network of knowledge' described by Kvan includes human and artificial agents alike, so long as they share knowledge and join in the problem-solving tasks. Along the same lines, Fischer (1999) also emphasises that the stakeholders who participate in design have equally limited knowledge that, when brought into a collaborative process, offers an opportunity for creativity. The notion of distribution is of paramount importance here, and must not be confused with the fact that the individual members of the group may be geographically far apart. Distribution implies that the nature of design is such that the knowledge and capabilities of the group do not derive simply from the summation of individual knowledge. Complex interactions developed between human and artificial members, adaptation and learning give rise to collective knowledge and 'cognitive properties that are not predictable from a knowledge of the properties of the individuals in the group' (Hutchins 1995: xiii).

The second dimension is that group design is increasingly perceived as a *social process*. In group design, interactions take place because complex *interdependencies* evolve between individual activities and decisions, which naturally give rise to *conflicts*. This epitomises the social character of group design: social behaviour develops from the need to resolve conflicts by taking part in a collaborative endeavour. Observations of design teams at work identified the occurrence of various forms of social interaction, such as negotiation, persuasion and compromise; and compliance with social norms, power structures, roles etc. (Branki *et al.* 1993; Cross and Cross 1995). However, it should again be emphasised here that, in reality, interactions evolve in – and between – human and artificial agent societies alike: computer models, tools and software agents are equally important carriers of individual knowledge and expertise. Therefore, our view of design as a social process should account for the fact that design takes place in a *socio-technical* context. Apart from the issues of conflict and social interaction, the social views of design also highlight another important issue. The recognition that the individual decision-makers come into the design process by contributing their expert (domain) knowledge also suggests that they consequently maintain different ways of understanding the design problem. Developing a common understanding of the problem (problem formulation/reformulation), and resolving conflicts that arise because of conflicting viewpoints and goals, is considered to be of primary importance if some shared outcome is to be achieved (Cross and Cross 1995; Kalay 1998). This also implies that individual and common objectives need to be simultaneously pursued and adapted.

The third dimension of group design decision-making is that *creativity is a collective attribute*. The question of creativity is the bone of contention between researchers in design studies who have been exploring questions related to the nature of creativity, the conditions that foster creative design, and the features of individual and social creativity. However, situating design within a social context means that both individual and collective creativity draws on collective knowledge, which is created by groups of people interacting with each other and with tools and artefacts (Fischer 1999). As mentioned above, collective design becomes in many cases a process of conflict resolution or compromise. However, this is not seen as an impediment to creativity; it may lead to thorough exploration or extension of the problem and solution spaces, and bring to light creative solutions that could not have been considered by an individual working in isolation.

Finally, it must be noted that effective collaboration in design is often linked with *communication*. Communication is generally acknowledged as an essential aspect of collaboration because it supports the exchange of information, knowledge and ideas, and facilitates detection and understanding of problems and conflicts. There are various arguments that strive to identify the best modes of communication: synchronous or asynchronous, mediated or face-to-face, etc. However, communication in the present chapter will be considered in a generic way, as a means to establish a common ground upon which social interactions are built. In this sense indirect and tacit modes of communication will be considered equally important. Such indirect forms of communication can be established, for example, with the use of external representations such as sketches and drawings. Concepts of indirect communication external to cognitive science research could also present mechanisms for the development of collaborative activity and coordination. For example, Susi and Ziemke (2001) argue that the concept of *stigmergy*, which explains how the behaviour of a group of social insects can be coordinated by indirect interaction (through modification of the physical environment), could offer valuable insights into the study of social interactions.

2.2 Comparing cooperation, collaboration and coordination

The four dimensions of group design elaborated above can help us elucidate the meaning of cooperation, coordination and collaboration, which come into view as three subtly – but crucially – different manifestations of multi-agent design. For example, Branki (1994) argues that groupwork involves all the three processes. Cooperation generally refers to a process of working together to develop a design artefact, while collaboration involves more specifically sharing, exchanging and maintaining information to create a common pool of knowledge and build shared goals. Coordination, on the other hand, is a concept more related to ordering of processes to ensure harmonious work, and may have four components: goals, activities, actors and interdependencies (ibid.: 37). Kvan (2000) offers a more comprehensive account of the three concepts. In his view, cooperation is a more low-level concept used to describe informal relationships between the members of a group and does not imply any commonly defined mission or goal. Collaboration, on the other hand, is seen as joint problem-solving and therefore requires a higher sense of working together and greater commitment to a common goal. In between, coordination is characterised by some formal relationships and understanding of compatible missions, but does not imply full commitment to common goals (Kvan 2000: 410–11). It follows that group design, although it can be defined as a social act, it is not necessarily collaborative.

The position adopted in the present study is that, in contrast to cooperation, coordination is an abstraction that matches well with the view of design as a social process but, unlike collaboration, is generic enough to capture a large class of design problems. The main issue captured by the concept of coordination is that decisions and actions of individuals that participate in a group (human or artificial agents) are interdependent. This necessarily leads to the development of some (weak) form of social structure that sets the scene for the effective management of interdependencies, resolution of conflicts, and construction of collective knowledge. Furthermore, the concept of coordination allows us to include, in our understanding of group design, processes where there are no shared goals, and collaboration is not a prerequisite for the formation of design solutions (see the concept of unstructured collaboration discussed by Craig and Zimring 2000).

3 Insights from artificial intelligence and multi-agent systems

In *The Sciences of the Artificial*, Simon (1969) established the view of design as a process that can be modelled and performed computationally, but more importantly he also established the view of design as the core of the sciences of the artificial. It is in this sense that devising intelligent machines becomes a tool for both the understanding and the devising of our world. In a similar vein, Winograd and Flores (1986: 4) perceive the question of design as 'the interaction between understanding and creation': a discourse between understanding our environment (and ourselves) and creating tools to re-interpret and change this environment.

Here we shall focus on a subfield of artificial intelligence (AI) which engages in the study of groups of intelligent agents, namely distributed artificial intelligence (DAI).

The general standpoint in DAI is that individual agents have limited knowledge and resources, information-processing abilities, and viewpoints, and therefore need to engage in some kind of collective activity or interaction in order to improve their problem-solving and goal-attainment capabilities. It goes without saying that the term 'distribution' is the quintessence of artificial societies: it suggests the existence of incomplete and dispersed information, interdependency of actions and decisions, and lack of global control mechanisms or rules to dictate global behaviour.

3.1 An overview of distributed artificial intelligence

Traditionally, DAI scientists were divided between two areas of research (Bond and Gasser 1988; Moulin and Chaib-Draa 1996): distributed problem-solving and multi-agent systems. In brief, *distributed problem-solving* (DPS) studies how a group of intelligent agents (or nodes) can share their resources and harmonise their activities, so that they can collectively solve a particular problem. According to Ossowski (1999: 39–40), DPS focuses on aspects of *cooperation* within a dynamic problem-solving environment, and typically conforms to three assumptions: the *benevolence* assumption, the *common goal* assumption, and the *homogeneous agents* assumption. This means that agents are in principle predisposed to cooperation and aware that they pursue a common goal, therefore conflicts only arise because of limited viewpoints and not because of conflicting goals or interests. On the other hand, research in *multi-agent systems* (MAS) is concerned with a loosely coupled network of agents who may work together to solve a problem. The focus therefore is on the *interactions* between agents, and the main assumptions are *non-benevolence, multiple goals, autonomy* and *heterogeneity* (Ossowski 1999: 43–4). In other words, individual agents are assumed to act based on individual rationality, which suggests that they maintain multiple, partially conflicting goals and interests. Despite their 'self-interested' nature, agents may wish to interact with each other so as to perform activities that assist them in the achievement of their own objectives (Lesser 1999). Heterogeneity suggests that other kinds of conflicts may also arise due to incompatible communication protocols, or agent architectures.

The MAS view offers a more generic perspective on distributed artificial intelligence and in recent years has been used in an inclusive sense to refer to the study, modelling and construction of artificial societies (for overviews, see Ferber 1999; Weiss 1999; and Wooldridge 2002).

3.2 Multi-agent systems in urban planning

In Urban Studies, multi-agent systems have been used mainly as a simulation tool. Multi-agent modelling in urban development and planning is a relatively new research area that has grown from the tradition of using artificial intelligence and artificial life techniques (mainly cellular automata) to investigate and model spatial dynamics such as land-use change, urban growth, population dynamics, traffic etc. (for overviews, see Besussi and Cecchini 1996; White and Engelen 2000). Adopters of the MAS paradigm have used agent technology to capture and represent the dynamics of individual behaviour in cities (e.g. Batty *et al.* 1998; Benenson 1998; Batty *et al.* 2002; Dijkstra and Timmermans 2002). These models have focused on the reproduction, visualisation and analysis of dynamic behaviour and have served mainly exploratory, descriptive and, to a certain extent, predictive purposes. So far, the majority of such models have not directly incorporated multiperson problem-solving and decision-making processes, therefore leaving out issues such as conflict resolution and goal adaptation; but also, more importantly for this research, they have not considered processes of collective plan generation. Some relevant research dealing with some of these issues is presented by Ligtenberg *et al.* (2001), Arentze and Timmermans (2002) and Saarlos *et al.* (2001).

3.3 Multi-agent systems in design

Research on MAS has been pursued within the 'AI in design' community to study design formally, and support it computationally, as a complex multidisciplinary process that involves group decision-making and collaborative working. Theoretical and technological advances in this area have gone hand in hand with advances in concurrent engineering, CSCW and collaborative CAD. Some groundwork research is reported in Gero and Maher (1993). Research on MAS in design can be roughly grouped under two main issues or directions of investigation: the development of models of (collaborative) design, and the development of design support systems.

Models of design may focus on the development of theories of collaborative design, the formalisation of design process, but also the definition and conceptualisation of design agents and their properties. Examples here include the A-Design theory proposed by Campbell *et al.* (1998, 1999), the studies presented by Brazier *et al.* (1996, 2000) and the ideas developed by Gero and his colleagues (e.g. Saunders and Gero 2001; Kannengiesser and Gero 2002; Gero and Sosa 2002). Gero and colleagues in particular have been focused on understanding design cognition of agents that are situated and interact with a certain environment; as well as modelling issues such as creativity, concept formation (Gero and Fujii 2000), and learning (Reffat and Gero 1999; Liew and Gero 2002).

Researchers interested in modelling design processes and agents have also focused on issues of conflict management and resolution (e.g. Klein and Lu 1989; Klein 1991; Brown *et al.* 1994; Dunskus *et al.* 1995; Berker and Brown 1996; Grecu and Brown 1996).

The second direction of investigation is the development of design decision support systems. This research is critically linked to the development of CSCW technologies, intelligent CAD systems, as well as models and simulations that can provide input into the decision-making process. Multi-agent technologies may be used for a range of purposes (and kinds of support) including the development of knowledge-rich systems, the design of (multi-) user interface(s), the integration of databases and tools, the development of languages for representation and communication, the creation of intelligent drafting and graphic recognition systems, and the development of systems for prediction and evaluation.

For example, Liu *et al.* (2002) see multi-agent technology mainly as a tool for combining diverse sources and types of information and reasoning, in order to support both human and artificial agents that take part in complex problem-solving processes in design. Lees *et al.* (2001) also discuss multi-agent technology for design support in the context of concurrent engineering. Edmonds *et al.* (1994) consider the use of multi-agent technology to provide for more creative aspects of design. They emphasise the importance of developing agents that can potentially perceive emergent forms in design (object) representations. Multi-agent recognition of graphic units, shapes or sketches is also used for the development of intelligent drafting tools (Achten 2002; Leclercq and Juchmes 2002). Finally, some researchers have experimented with multi-agent models, such as swarms, to explore and/or generate architectural spaces and forms (Küppers *et al.* 2000; Coates and Thum 2000).

Another class of applications, developed for decision support, employ multi-agent simulations to explore and visualise individual behaviour as a means of evaluating designs. For example, Dijkstra and Timmermans (2002) describe a multi-agent model, developed by combining multi-agent technology and cellular automata (CA), for the simulation of individuals moving within buildings such as shopping malls or public places. Multi-agent simulation in this case is used to predict and analyse pedestrian traffic, in order to assess the outcome of design choices and decisions. Saunders and Gero (2002) describe another model developed specifically to support the evaluation of environments that are designed to stimulate exploration, such as galleries and museums. Note that in both examples coordination is achieved essentially through reactive-like rules.

4 The basic dimensions of multi-agent design as coordination

This chapter started with the recognition that coordination is inherent in design and planning processes evolving in complex, dynamically changing, multi-agent environments. The recognition of coordination as a crucial problem in design and planning is also a recognition of the multi-agent nature of these processes.

Let us discuss some critical aspects of multi-agent design related to coordination. At the most fundamental level, coordination is associated with distribution. The distribution of decision-making in multi-agent settings suggests that there is no central authority capable of controlling the process and outcome of design. In this sense, coordination is related to distributed control. An important corollary of this observation is that knowledge is also distributed among individual agents (human or artificial) that take part in the process. In this sense, sharing of information and learning are very desirable for improving the adaptability of agents in their ever-changing environment, and for ensuring consistent problem-solving behaviour. As these conditions can be associated with multi-agent coordination in general, it is also important to highlight how these conditions are understood and further adapted in the context of design and planning.

In design, the articulation and pursuit of goals is not only a collective process but also a process that binds together problem-finding and problem-solving. In this sense, adaptation of goals and expectations, generation and evaluation of alternatives, and co-evolution of problem and solution are all essential characteristics of coordination in design domains. The recognition of these characteristics is important for understanding coordination not only in relation to the management of activities (i.e. ordering, scheduling and synchronisation), which is common knowledge in engineering domains, but also in relation to the generation and adaptation of knowledge and decisions.

Moreover, design and planning are social processes that involve dealing with conflict that arises from incompatible viewpoints and goals. This is followed by a need to construct some form of common understanding (of the problem) through adaptation of goals and expectations. Learning is an important vehicle for the creation of collective knowledge and common understanding. Finally, understanding design and planning as social processes implies that distinctive abilities associated with design, such as creativity, are (and should be) perceived as collective properties. To understand and potentially model coordination in design domains necessitates therefore understanding and formalising creativity as a collective process.

The position of this chapter is that coordination can be used as an abstraction for the definition of design as a social process, wherein interdependency, conflict and construction of collective knowledge become the focal points. Additionally, in this setting, coordination can be linked to creativity through the concept of emergence, which involves considering the relation and interaction between the local level of individual agents, and the global level of the collective entity (social group).

We can now summarise the crucial dimensions of multi-agent design using the abstraction of coordination as follows:

- First, distributed design tasks require knowledge that is spread among local agents. Coordination thus involves synthesising or constructing the knowledge necessary for the collective task. In this sense, learning is seen as an important instrument not only for enhancing the individual ability of agents to derive design solutions, but also for creating shared knowledge about the design task and its constraints.
- Second, in multi-agent design, decisions are driven by individual goals and requirements, and are taken at a local level without any external centralised source of control. Coordination is thus seen as a distributed control process which leads to the emergence of design solutions.
- Finally, in distributed design decision-making, the definition of problem and solution spaces also becomes a collective undertaking. Coordination signifies the need for a parallel exploration, generation, and reformulation of problem and solution spaces.

5 Modelling coordination

This section demonstrates how computer simulations can be used for theory development; in particular as a means to think through the dimensions and characteristics of coordination, and construct a coherent framework for its understanding. It is informative to consider this statement in more detail. The exploration above led to the development of a theoretical view of multi-agent design as coordination, captured by the conditions of learning, distributed control and co-evolution. These conditions are quite abstract, and instantiating them within a model provides the means to make them specific. The formalisation of the model within a simulation program in turn provides a way to check the consistency of the theory. As Gilbert and Troitzsch (1999: 5) note, 'the process of formalisation, which involves being precise about what the theory means and making sure that it is complete and coherent, is a very valuable discipline in its own right'.

5.1 Gero's FBS framework for modelling the design process

To start modelling coordination and exemplify its dimensions here, we utilise and adapt the function–behaviour–structure (FBS) framework proposed by Gero (1990, 2000). The framework is used because it offers a way to discuss design in a domain-independent way. In Gero's FBS framework, function is generally tied to the purpose of an artefact; structure relates to the form and organisation of the components of an artefact; and behaviour refers to the operation of an artefact through which a particular function is satisfied, or to some observation or measurement applied over a particular structure. Gero distinguishes two types of behaviour: structural behaviour, which is directly derived from structure; and expected behaviour, which is derived from function. Gero identifies a set of processes by which a designer moves from function to structure and thereby to an appropriate design description.

More analytically, these processes are: (*a*) *analysis*, where the behaviour of a structure Bs is deduced from this structure S; (*b*) *formulation*, where function F is mapped to expected behaviour Be; (*c*) *synthesis*, where the expected behaviour Be is used to produce a structure S, based on knowledge of the behaviours Bs; (*d*) *evaluation*, where the structural behaviour Bs is compared with the expected behaviour Be to determine whether the synthesised structure can satisfy the desired function, (*e*) *reformulation*, where the range of expected behaviours Be can be changed, and through them the function, and (*f*) design *description*, where the structure is developed into a description of the artefact to be constructed. In later papers (e.g. Gero and Kannengiesser 2002), two other types of reformulation are also added: one which refers to changes in terms of structure variables or ranges of values for them, and one which refers to changes in terms of function variables or ranges of values of them. For a summary, see Table 5.1.

Table 5.1 Summary of the processes involved in design as discussed by Gero and Kannengiesser (2002)

Analysis:	S → Bs
Formulation:	F → Be
Synthesis:	Be → S
Evaluation:	Be ↔ Bs
Reformulation:	S → Be
	S → S
	S → F
Design description:	S → D

5.2 Employing the FBS framework to propose a model of coordination

From this presentation, the processes of analysis and synthesis can be taken to represent operations that concern the solution space (the space of possible structures), whereas formulation and reformulation concern the problem space (the space of possible functions). This gives a means to understand and model how a co-evolution of problem and solution spaces may be realised. The conceptual model of coordination introduced here develops this idea and recasts the processes of analysis, synthesis, formulation, reformulation and evaluation within the context of distributed learning control.

More analytically, a distributed design task involves a number of agents that act on a common world while trying to achieve their individual visions about this world. Thus there is a direct interaction between an agent and its environment, which includes other agents. Each agent is considered to carry out two combined control-based activities depicted in Figure 5.1.

The first control activity alludes to an analysis–synthesis–evaluation route. The objective of each agent in this case is to find a suitable path or class of structures S (control actions) that can lead the behaviours Bs to follow the expected behaviour Be (derived by a target function F), despite uncertainties and despite exogenous disturbances Sd produced by other agents' decisions. Here, knowledge of associations between proposed structures S and actual behaviours Bs, and the observed distance or 'error' between expected and structural behaviours are both used to inform the selection of suitable control actions S.

The second activity alludes to an evaluation–formulation–reformulation route. The objective of each agent in this case is to find a suitable function F leading to such expected behaviour Be that can satisfy a given structural configuration S, despite uncertainties and despite exogenous disturbances Fd (i.e. other agents' goals). Here, again, knowledge of associations between proposed functions F and the overall derived behaviour Be, as well as the 'error' between this and the actual behaviour of the world, is used to inform the generation of suitable targets F.

In effect, the two control processes provide targets and constraints for one another: the desired performance of the analysis–synthesis process is evaluated

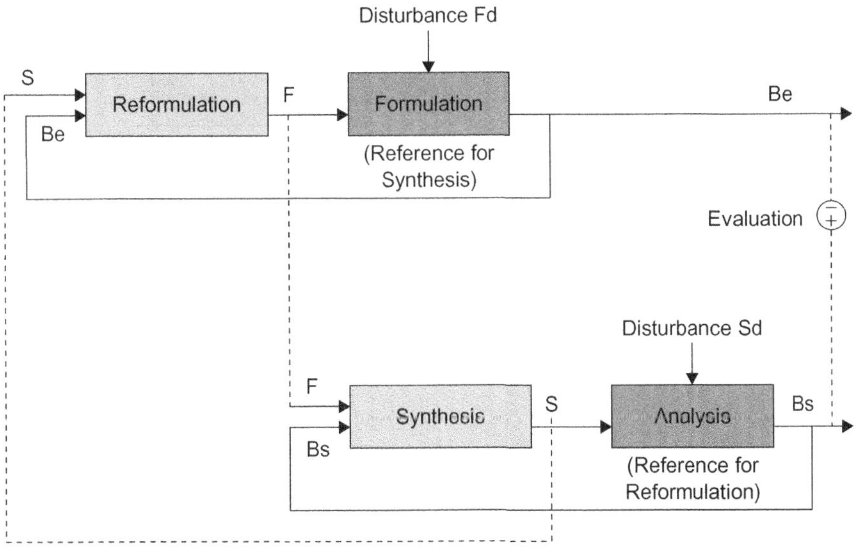

Figure 5.1 Coordination as a dual control process, one that corresponds to a synthesis–analysis–evaluation route and one that corresponds to a reformulation–formulation–evaluation route

(denoted by E in the figure) through the formulation–reformulation process and vice versa. Overall, each agent uses this dual-control process in order to lead the common world to perform according to their individual goals.

This conceptualisation recognises that each agent's goals and solutions may be conflicting and explicitly makes this part of the design process. Distributed control captures the idea that each agent needs to arrive at design solutions by avoiding conflicts (or disturbances) introduced into the world through the action of other agents. The process of solution generation based on learning control is a process of self-adaptation of agents that leads to coordination of their distributed descriptions.

5.3 Building the distributed learning control model

5.3.1 The design problem and the representation of functions, behaviours and structures

To experiment with this conceptual model, we take the problem of devising the layout of a building. Let us consider a design situation involving three agents, each with a different (initial) goal for the configuration of the building plan.

The space where the actions of the agents are expressed was realised by using the representation proposed by Steadman *et al.* (Steadman 1998; Steadman and Waddoups 2000). The 'archetypal building' is an abstract representation of a large class of rectangular built forms. This abstract form is composed from three types of spaces defined according to the type of lighting they incorporate. The archetype therefore contains spaces that are naturally lit, spaces that are artificially lit (like corridors) and spaces lit from the top (topmost courtyard floor and courtyards).

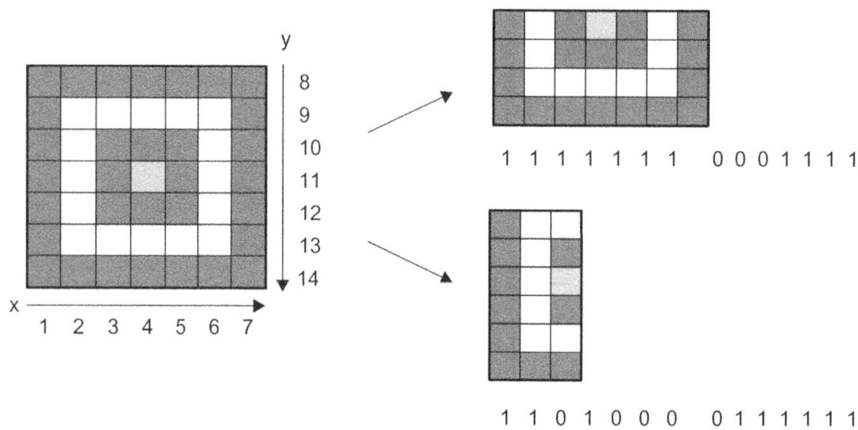

Figure 5.2 The binary encoding of the archetypal building proposed by Steadman and Waddoups (2000). The fourteen-digit binary array can be used to describe the geometry of rectangular built forms – as shown, for example, on the right (where 1 represents the presence and 0 the absence of a strip in the *x* and *y* axes). Here white refers to artificially lit space, dark grey to side-lit space, and light grey to top-lit space

Steadman *et al.* showed that from this abstract form, and by using a series of transformations (basically subtraction), a very great variety of rectangular forms can be produced, corresponding to different types of buildings found in the real world. The nucleus of the archetype is a subset of the overall form and contains one courtyard (Figure 5.2). The authors suggested an encoding of this form by using a fourteen-digit binary string, of which the first seven digits represent positions in a horizontal axis (*x* axis) and the remaining seven digits represent positions in a vertical axis (*y* axis). By assigning a value of either 0 or 1 in any such position, one can add and subtract strips to create different forms (see Figure 5.2).

The representation adopted for the experiment described here (Figure 5.3) is slightly different from the original in that counting of the *y* positions (or strings) starts from the bottom. Also in this experiment, each position is not defined by a binary value, but instead takes a value in the range between 0 and 1, so that a metric (geometrical) space can be created. The representation used is two-dimensional, and we consider here a single-storey building layout, although one could envisage assigning a third dimension along an orthogonal axis *z* to represent height.

The archetype is used as a space of possibilities within which agents' proposals can be developed. Here functional and structural features are captured

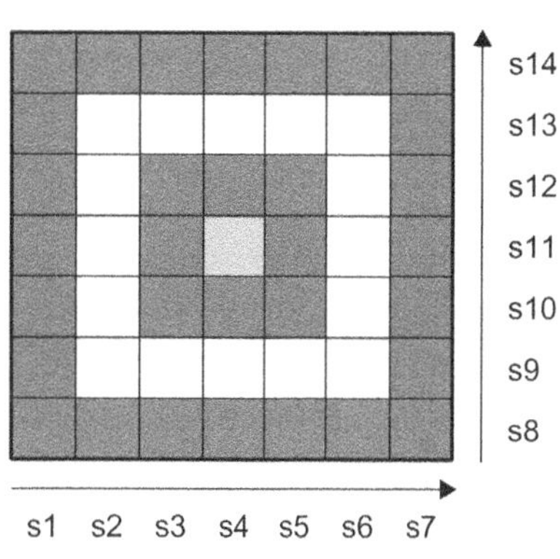

Figure 5.3 The representation of the archetype used in the experiment, adapted from Steadman and Waddoups (2000)

within a single representation as the structural (topological) dimensions are defined according to their function (lighting). Behavioural information is captured as change of functional and structural characteristics in order to achieve individual goals.

Goals in this model are represented directly in terms of archetype configurations. The idea is that different configurations express in principle different preferences about the organisation of space in relation to lighting, but these preferences are not explicitly articulated. This allows us to express situations in which initial ideas about architectural solutions come in the form of rough plan layouts. The knowledge behind the process of formulating and reformulating goals is tacit, and distributed learning control here can be thought as a kind of 'design reasoning without explanations' (see Coyne 1990). The downside of this representation of goals is that it makes the semantics of the design object implicit, and seems to place a somewhat unbalanced emphasis on the manipulation of abstract forms as the core of the design process.

5.3.2 The control architecture

The model was implemented in MATLAB. The control architecture adopted uses two neural networks: one to act as a controller (a system that controls) and another to act as the world model (a model of the space where design solutions are developed). Additionally, the world model is also used as a reference model to provide targets for the formulation–reformulation process. The control architecture is pictured in Figure 5.4. The architecture represents the workings of each of the three agents.

More specifically, the experiment is set up as follows. The three agents interact with each other in a sequential way (Figure 5.5). Each agent uses the first neural network (NN Model) to learn input–output pairs of descriptions, that is, associations between its own proposals (control actions) directed to the next agent and the replies from the agent two steps down the line. So what constitutes the observable world for one agent is a composite of the two other agents. Namely, the world model of Agent1 is formed by learning associations between u1 and u2. Similarly, the world model of Agent2 is formed by associations (u2, u3) and the world model of Agent3 by associations

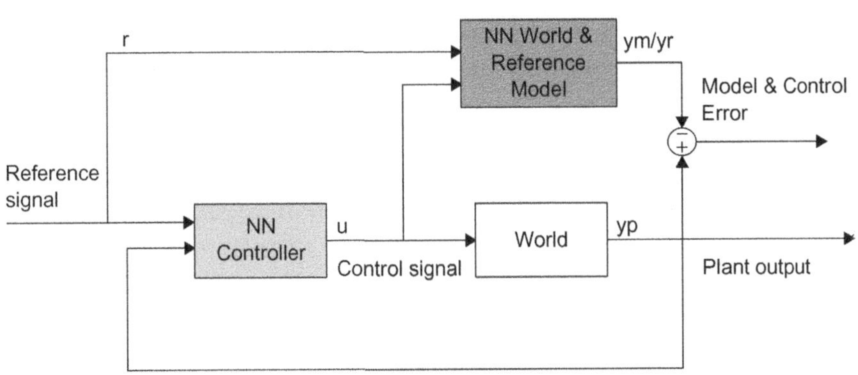

Figure 5.4 The control architecture used in the experiment, which represents the workings of each agent. The world model is also used as a reference model

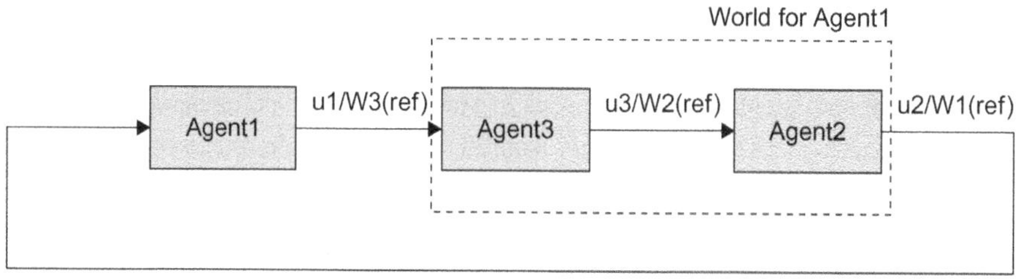

Figure 5.5
An illustration of the sequential connection between agents. The world for one agent is a composite of the other two

(u3, u1). The second neural network which acts as a controller essentially learns the inverse of the world model. The inputs for each agent are presented as fourteen-digit arrays corresponding to geometrical configurations (instances of the archetypal building). The general idea, then, is that agents observe and learn each other's behaviours, and use this knowledge to propose configurations.

5.4 *Results and reflections*

Figure 5.6 shows results from a simulation run for time t = 100. The control actions of the three agents are displayed in a sequence: u1, u2, u3. The vertical succession denotes time (simulation cycles). In the first cycle (t = 1) the control actions result from the simulation of each agent's controller given an initial world W (Figure 5.7) and three random inputs – one for each agent. In the second cycle (t = 2), the results are obtained after the controller is trained for the first time using the control actions and targets/expectations from the previous cycle. Owing to the control architecture, in all subsequent iterations the topology of the control action of each agent remains the same, although the specific values change. Compare, for example, the control actions for t = 2 and t = 37 in the figure. From t = 37 to t = 100 also the geometry (the values) of the control actions remains the same for all agents. This precisely reflects the control algorithm as described above, which determines how the control actions of one agent become targets of another.

There is a very important reason for this dead-end situation. The setting of the experiment equates the world at any given moment with the control action of each agent active at that moment. Hence, the world is not a common space where the control actions (and design proposals) of agents can be collectively expressed with all their mutual or conflicting effects. In particular, the notion of disturbance introduced in the conceptual model at the beginning of this chapter is completely eliminated. In this sense the task of creating a world model (a model of each agent's actions) is duly straightforward. The role of disturbance was overlooked in the experiment. Although disturbance is a source of noise in the face of which agents have to learn to plan their control actions, it is also an important source of variation. Coordination is significantly reliant on the existence of disturbance, *a direct effect of the fact that design is distributed.* Other agents may be causes of conflict, but may also bring new knowledge, new

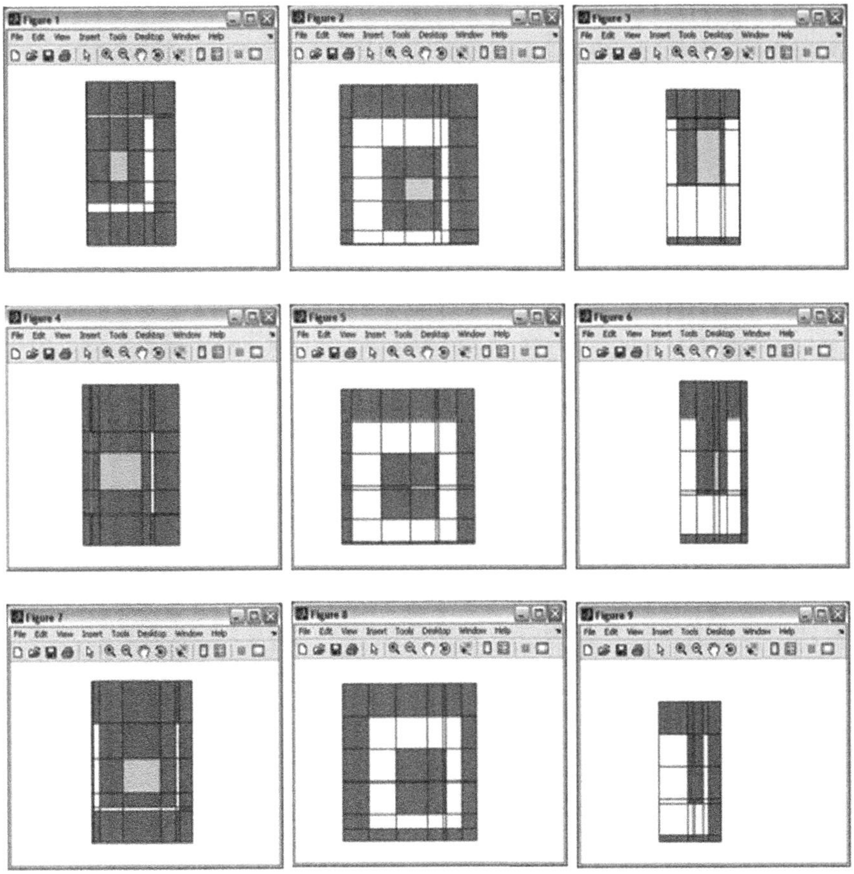

Figure 5.6 Results from a simulation run for time t = 100. The figures illustrate the control actions of the three agents, given an initial world (Figure 5.7), and for times t = 1, t = 2 and t = 37, from top to bottom

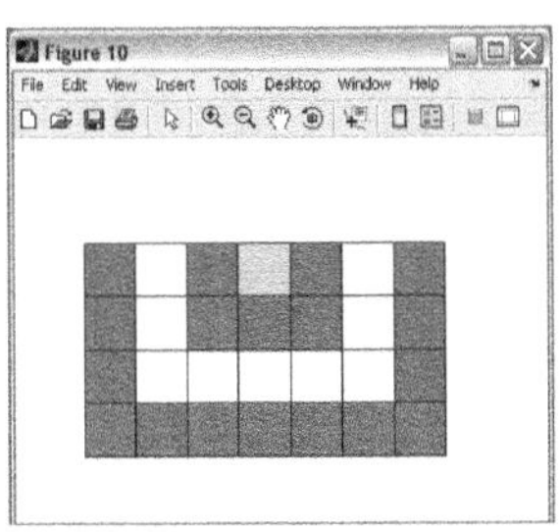

Figure 5.7 The initial world provided to the three agents for the simulation

resources, and new opportunities for creativity. In the model of coordination proposed here, disturbance works to an extent as a counterpart of the idea of random variation which we encounter in evolutionary and co-evolutionary models of design.

Additionally, by conflating the world model with the reference model in the control architecture the agents are forced to adopt as their goals the control actions of others. In effect, agents simply exchange goals, and no progress with the design coordination process can be made.

The omission of the discrepancy between the reference model and the world model is linked to a failure to distinguish between expectations and goals, and is more deeply rooted in a limitation of the FBS framework used. As described previously (see, for example, Table 5.1), expected behaviours in this framework are considered to be behaviours derived from function and are therefore uniquely tied to goals. There is no account for expected behaviour as behaviour derived from a model of the world, or knowledge about the effects of other agents' actions. Indeed, these two are different, and reducing one to the other in the simulation was a crucial flaw.

6 Revisiting the model of coordination as distributed learning control

Taking into consideration the arguments explored above, it seems necessary to revise the conceptual model presented at the beginning of this chapter with a more appropriate diagrammatic representation, where the difference between expected and desired behaviours is made explicit (Figure 5.8). The new representation also makes clear the idea that the analysis and formulation components incorporate the 'world', the place where the actions of agents are expressed to generate designs.

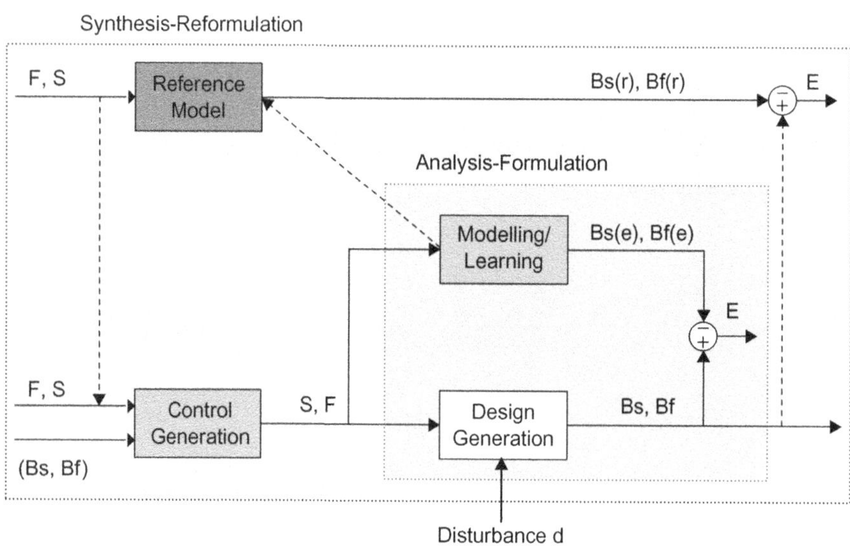

Figure 5.8 The updated model of coordination as distributed learning control. F = Function, S = Structure, Bf = Behaviour derived from function, Bs = Behaviour derived from structure, Bf(e) and Bs(e) = Expected Behaviours, Bf(r) and Bs(r) = Desired Behaviours, E = Evaluation

In this coordination model, therefore, expectations express knowledge about future states of the world and are formed on the basis of observations of current and past states of this world. The acquired knowledge is used to guide the generation and selection of control actions able to satisfy given goals or desires. The reference model is also created on the basis of knowledge about the world and observations about the consequences or effectiveness of actions. But, while expectations are used to infer control actions, goals are used as a way to evaluate the boundaries of the control actions or, in other words, to delineate desired possibilities for the world that is being designed. As discussed, the creation of goals, in parallel to the creation of design solutions, is a very vital requirement for creative design in general and coordination in particular, where these processes are distributed. We should keep in mind that goals in this model are formed both for the synthesis (solution formation) and the reformulation (problem formation) processes. What seems to be especially important, however, is not only to understand and model the transformation of goals in the distributed design process, but also to give an account of how the function that evaluates these goals is modified in the process (i.e. the mechanism that decides when and how these goals should be transformed).

In the model of coordination proposed here, the reference model, which is responsible for the transformation and evaluation of goals, is developed on the basis of a kind of social learning. That is, the knowledge of agents is formed through observations of the world within which their own as well as others' actions are expressed. Although neural networks were employed in the experiments, other types of learning mechanisms could be appropriate and are worthy of exploration. The question of how the social context may influence agents' decisions and goals is crucial, and considering alternatives for expressing this idea would be valuable. For example, an interesting experiment would be to include a representation of the social network of agents (a representation of the way agents are connected) and develop a process such that this network influences the formation of agents' knowledge and goals, and (possibly) vice versa. This experimentation would be useful in exploring various hypotheses about successful modes of social learning, successful strategies for goal adoption, or factors for achieving better coordination in different cases.

7 Conclusion and discussion

The proposition of using the concept of coordination as an abstraction of multi-agent design comes with a number of benefits. At the core of the notion of coordination is the idea that the decisions and actions of agents in a distributed design setting are interdependent. Coordination therefore places emphasis on distribution, interdependence and complexity, and facilitates the understanding of multi-agent design as a social process that includes collaboration as well as conflict. Although there is a growing body of studies interested in social aspects of design, these are usually concerned with descriptions of work practices, or of interactions between designers within a particular setting, and do not provide a general theory of multi-agent design which can be used to observe and interpret reality. The dimensions of multi-agent design identified in the chapter (i.e. the dimensions of learning, decentralised control and co-evolution) are proposed as fundamental features of coordination which can be used for observing, interpreting and even facilitating design activity. In the research reported here, computational models and simulations provide a framework so that the main assumptions of the model can become concrete and manifested in a coherent way. The experimentation helped reconsider and refine the theoretical ideas leading to a better understanding of design.

References

Achten, H. (2002), 'An approach for agent-systems in design support', in *Intelligent Agents in Design Workshop notes, AID '02 Conference*, Cambridge.

Arentze, T. and Timmermans, H. (2002), 'A multi-agent model of negotiation processes between multiple actors in urban developments: a framework and results of numerical experiments', unpublished paper, Eindhoven University of Technology.

Batty, M., Desyllas, J. and Duxbury, E. (2002), 'The discrete dynamics of small-scale spatial events: agent-based models of mobility in carnivals and street parades', CASA Working Paper 56. University College London, Centre for Advanced Spatial Analysis (CASA).

Batty, M., Jiang, B. and Thurstain-Goodwin, M. (1998), 'Local movement: agent-based models of pedestrian flows', CASA Working Paper 4. University College London, Centre for Advanced Spatial Analysis (CASA).

Benenson, I. (1998), 'Multi-agent simulations of residential dynamics in the city', *Computers, Environment and Urban Systems*, 22 (1): 25–42.

Berker, I. and Brown, D. C. (1996), 'Conflicts and negotiation in single function agent based design systems', *Concurrent Engineering: Research and Applications, Special Issue on Multi-Agent Systems in Concurrent Engineering*, ed. D. C. Brown, S. Lander and C. Petrie, 4 (1): 17–33.

Besussi, E. and Cecchini, A. (eds) (1996), *Artificial Worlds and Urban Studies*, Venice: DAEST, IUAV.

Bond, A. H. and Gasser, L. (eds) (1988), *Readings in Distributed Artificial Intelligence*, San Mateo, Calif.: Morgan Kaufmann.

Branki, N. E. (1994), 'An AI-based framework for collaborative design', in J. S. Gero, M. L. Maher and F. Sudweeks (co-chairs, eds), *Papers from the AAAI Workshop: AI and Collaborative Design* AAAI Press, pp. 35–46.

Branki, N. E., Edmonds, E. A. and Jones, R. M. (1993), 'A study of socially shared cognition in design', *Environment and Planning B: Planning and Design*, 20 (3): 295–306.

Brazier, F. M. T., Jonker, C. M. and Treur, J. (1996), 'Modelling project coordination in a multi-agent framework', *Proceedings of the Fifth Workshop on Enabling Technologies: Infrastructure and Collaborative Enterprises (WET ICE '96)*, Los Alamitos, Calif.: IEEE, pp. 148–55.

Brazier, F. M. T., Jonker, C. M. and Treur, J. (2000), 'Compositional design and reuse of a generic agent model', *Applied Artificial Intelligence*, 14 (5): 491–538.

Brown, D. C., Dunskus, B. and Grecu, D. (1994), 'Using single function agents to investigate negotiation', in M. Klein and S. Lander (eds), *AAAI Workshop: Models of Conflict Management in Cooperative Problem Solving*, Los Alamitos, Calif.: AAAI Press, pp. 19–23.

Campbell, M. I., Cagan, J. and Kotovsky, K. (1998), 'A-design: theory and implementation of an adaptive, agent-based method for conceptual design', in J. S. Gero and

F. Sudweeks (eds), *Artificial Intelligence in Design '98 (AID '98)*, Dordrecht, Kluwer, pp. 579–98.

Campbell, M. I., Cagan, J. and Kotovsky, K. (1999), 'A-design: an agent-based approach to conceptual design in a dynamic environment', *Research in Engineering Design*, 11 (3): 172–92.

Coates, P. and Thum, R. (2000), 'Parallel systems and architectural form', in *Proceedings of Greenwich 2000: Digital Creativity Symposium*, Greenwich: University of Greenwich, pp. 101–9.

Coyne, R. D. (1990), 'Design reasoning without explanations', *AI magazine*, 11 (4): 72–80.

Craig, D. L. and Zimring, C. (2000), 'Supporting collaborative design groups as design communities', *Design Studies*, 21 (2): 187–204.

Cross, N. and Cross, A. C. (1995), 'Observations of teamwork and social processes in design', *Design Studies*, 16 (2): 143–70.

Dijkstra, J., Jessurun, J. and Timmermans, H. J. P. (2002), 'A multi-agent cellular automata model of pedestrian movement', in M. Schreckenber and S. D. Sharma (eds), *Pedestrian and Evacuation Dynamics*, Berlin: Springer-Verlag, pp. 173–80.

Dijkstra, J. and Timmermans, H. (2002), 'Towards a multi-agent model for visualizing simulated user behavior to support the assessment of design performance', *Automation in Construction*, 11 (2): 135–45.

Domeshek, E. A., Kolodner, J. L., Billington, R. and Zimring, C. M. (1994), 'Using theories to overcome social obstacles in design collaborations', in J. S. Gero, M. L. Maher and F. Sudweeks (co-chairs, eds), *Papers from the AAAI Workshop: AI and Collaborative Design*, Los Alamitos, Calif.: AAAI Press, pp. 143–8.

Dorst, K. and Cross, N. (2001), 'Creativity in the design process: co-evolution of problem-solution', *Design Studies*, 22 (5): 425–37.

Dunskus, B. V., Grecu, D. L., Brown, D. C. and Berker, I. (1995), 'Using single function agents to investigate conflicts', *Artificial Intelligence in Engineering Design, Analysis and Manufacturing (AIEDAM), Special Issue: Conflict Management in Design*, ed. I. Smith, 9 (4): 299–312.

Edmonds, E. A., Candy, L., Jones, R. and Soufi, B. (1994), 'Support for collaborative design: agents and emergence', *Communications of the ACM*, 37 (7): 41–7.

Ferber, J. (1999), *Multi-Agent Systems: An Introduction to Distributed Artificial Intelligence*, London: Addison-Wesley.

Fischer, G. (1999), 'Social creativity, symmetry of ignorance and meta-design', in *Proceedings of the Third Conference on Creativity and Cognition*, Loughborough: ACM, pp. 115–23.

Gero, J. S. (1990), 'Design prototypes: a knowledge representation schema for design', *AI Magazine*, 11 (4): 26–36.

Gero, J. S. (2000), 'Computational models of innovative and creative design processes', *Technological Forecasting and Social Change*, 64 (2–3): 183–96.

Gero, J. S. and Fujii, H. (2000), 'A computational framework for concept formation for a situated design agent', *Knowledge-Based Systems*, 13 (6): 361–8.

Gero, J. S. and Kannengiesser, U. (2002), 'The situated function–behaviour–structure framework', in J. S. Gero (ed.), *Artificial Intelligence in Design '02*, Dordrecht: Kluwer, pp. 89–104.

Gero, J. S. and Maher, M. L. (1993), 'Artificial intelligence in collaborative design', Technical Report WS-93-07, Menlo Park, Calif.: AAAI Press.

Gero, J. S. and Sosa, R. (2002), 'Creative design situations: artificial creativity in communities of design agents', in A. Eshaq, C. Khong, K. Neo, M. Neo and S. Ahmad (eds), *CAADRIA 2002*, New York: Prentice Hall, pp. 191–8.

Gilbert, N. and Troitzsch, K. G. (1999), *Simulation for the Social Scientist*, Buckingham: Open University Press.

Grecu, D. L. and Brown, D. C. (1996), 'Learning by single function agents during spring design', in J. S. Gero and F. Sudweeks (eds), *Artificial Intelligence in Design '96 (AID '96)*, Boston, Mass.: Kluwer, pp. 409–28.

Hutchins, E. (1995), *Cognition in the Wild*, Cambridge, Mass.: The MIT Press.

Kalay, Y. E. (1998), 'P3: computational environment to support design collaboration', *Automation in Construction*, 8 (1): 37–48.

Kannengiesser, U. and Gero, J. S. (2002), 'Situated agent communication for design', in J. S. Gero and F. M. T. Brazier (eds), *Agents in Design 2002*, Sydney: Key Centre for Design Computing and Cognition, University of Sydney, pp. 85–94.

Klein, M. (1991), 'Supporting conflict resolution in cooperative design systems', *IEEE Transactions on Systems, Man and Cybernetics*, 21 (6): 1379–90.

Klein, M. and Lu, S. C. Y. (1989), 'Conflict resolution in cooperative design', *Artificial Intelligence in Engineering*, 4 (4): 168–80.

Küppers, S., Gage, S., Penn, A. and Mottram, C. (2000), 'Distributed simulation system for spatial design: parallel physical and virtual interactions between people and objects', in *Proceedings of Greenwich 2000: Digital Creativity Symposium*, Greenwich: University of Greenwich, pp. 87–99.

Kvan, T. (1999), 'Designing together apart: computer supported collaborative design in architecture', in *Department of Design and Innovation*, Milton Keynes: Open University.

Kvan, T. (2000), 'Collaborative design: what is it?', *Automation in Construction*, 9 (4): 409–15.

Leclercq, P. and Juchmes, R. (2002), 'The absent interface in design engineering', *Artificial Intelligence in Engineering Design, Analysis and Manufacturing (AIEDAM), Special Issue: Human–Computer Interaction in Engineering Contexts*, ed. I. C. Parmee and I. F. C. Smith, 16 (3).

Lees, B., Branki, C. and Aird, I. (2001), 'A framework for distributed agent-based engineering design support', *Automation in Construction*, 10 (5): 631–7.

Lesser, V. R. (1999), 'Cooperative multiagent systems: a personal view of the state of the art', *IEEE Transactions on Knowledge and Data Engineering*, 11 (1): 133–42.

Liew, P.-S. and Gero, J. S. (2002), 'An implementation model of constructive memory for a design agent', in J. S. Gero and F. M. T. Brazier (eds), *Agents in Design 2002*, Sydney: Key Centre for Design Computing and Cognition, University of Sydney, pp. 275–6.

Ligtenberg, A., Bregt, A. K. and Van Lammeren, R. (2001), 'Multi-actor-based land use modelling: spatial planning using agents', *Landscape and Urban Planning*, 56 (1): 21–33.

Liu, H., Tang, M. and Frazer, J. H. (2002), 'Supporting evolution in a multi-agent cooperative design environment', *Advances in Engineering Software*, 33 (6): 319–28.

Maher, M. L. (1994), 'Creative design using a genetic algorithm', *Computing in Civil Engineering*, 2014–21.

Moulin, B. and Chaib-Draa, B. (1996), 'An overview of distributed artificial intelligence', in G. M. P. O'Hare and N. R. Jennings (eds), *Foundations of Distributed Artificial Intelligence*, New York: John Wiley.

Ossowski, S. (1999), *Co-ordination in Artificial Agent Societies*, Berlin/Heidelberg: Springer-Verlag.

Reffat, R. M. and Gero, J. S. (1999), 'Situatedness: a new dimension for learning systems in design', in A. Brown, M. Knight and P. Berridge (eds), in *Architectural Computing: From Turing to 2000 (eCAADe '99)*, Liverpool: University of Liverpool, pp. 252–61.

Rittel, H. W. J. and Webber, M. M. (1984), 'Planning problems are wicked problems', in N. Cross (ed.), *Developments in Design Methodology*, New York, John Wiley, pp. 135–44.

Rosenman, M. A. and Gero, J. S. (1998), 'Purpose and function in design: from the socio-cultural to the techno-physical', *Design Studies*, 19 (2): 161–86.

Saarlos, D. J. M., Arentze, T. A., Borgers, A. W. J. and Timmermans, H. J. P. (2001), 'Introducing multi-agents in an integrated GIS/VR system for supporting urban planning and design decisions, in *Proceedings of the 7th International Conference on Computers in Urban Planning and Urban Management CUPUM 2001* (CD-ROM), Honolulu: University of Hawaii.

Saunders, R. and Gero, J. S. (2001), 'Artificial creativity: a synthetic approach to the study of creative behaviour', in J. S. Gero and M. L. Maher (eds), *Computational and Cognitive Models of Creative Design V*, Sydney: Key Centre for Design Computing and Cognition, University of Sydney, pp. 113–39.

Saunders, R. and Gero, J. S. (2002), 'Curious agents and situated design evaluations', in J. S. Gero and F. M. T. Brazier (eds), *Agents in Design 2002*, Sydney: Key Centre for Design Computing and Cognition, University of Sydney, pp. 133–49.

Simon, H. A. (1969), *The Sciences of the Artificial*, Cambridge, Mass.: The MIT Press.

Smithers, T. (1998), 'Towards a knowledge level theory of design process', in J. S. Gero and F. Sudweeks (eds), *Artificial Intelligence in Design '98*, Dordrecht: Kluwer.

Steadman, J. P. (1998), 'Sketch for an archetypal building', *Environment and Planning B: Planning and Design*, 27 (Anniversary Issue): 92–105.

Steadman, J. P. and Waddoups, L. (2000), 'A catalogue of built forms using a binary representation', in *Proceedings of the 5th International Conference on Design and Decision Support Systems in Architecture (DDSS)*, Einhoven University, pp. 353–73.

Susi, T. and Ziemke, T. (2001), 'Social cognition, artefacts and stigmergy: a comparative analysis of theoretical frameworks for the understanding of artefact-mediated collaborative activity', *Journal of Cognitive Systems Research*, 2 (4): 273–90.

Weiss, G. (ed.) (1999), *Multiagent Systems: A Modern Approach to Distributed Artificial Intelligence*, Cambridge, Mass.: The MIT Press.

White, R. and Engelen, G. (2000), 'High-resolution integrated modelling of the spatial dynamics of urban and regional systems', *Computers, Environment and Urban Systems*, 24 (5): 383–400.

Winograd, T. and Flores, F. (1986), *Understanding Computers and Cognition: A New Foundation of Design*, Norwood, NJ: Ablex.

Wooldridge, M. J. (2002), *An Introduction to MultiAgent Systems*, London: John Wiley.

This page intentionally left blank

6

The mathematical conditions of design ability

A complexity theoretic view

Theodore Zamenopoulos

1 Introduction

Much of the world we live in is the result of some sort of 'design activity'. It is the result of the capacity to form responses within a 'task' environment by recognising opportunities, formulating goals or visions, and plan the creation of entities that should satisfy these goals. For instance, the capacity of birds to build nests, beavers to build dams, or humans to construct hunting tools can be perceived as primitive examples of design. This capacity can be equally distributed within a society. For instance, in the creation of ant colonies and nests, or the creation of cities, the capacity to design is distributed among a society (a collective) of agents. Design is a fundamental aspect of human societies, manifested in their organisation and governance, as well as in their products, or the abstract and physical constructs created to support their functioning. According to this perspective, much of the world around us is essentially artificial; it is created, recognised and used by intentional organisms as a functional artefact. Based on these observations, design is treated here very broadly, as a general natural phenomenon that underlies the creation of artificial worlds. Design is seen as a capacity of individual or social organisms to recognise design artefacts and carry out design tasks. The main theme of this chapter is the identification of the conditions that underlie this capacity. We hope that this study will contribute both in understanding some fundamental dimensions of design-capable organisms, and also in understanding and supporting their activities.

However, design is difficult to study and define as it has many expressions; it is a biological, cognitive, social, but also cultural phenomenon. As a cognitive pheonomenon, it is associated with complex self-organising neurological processes. As a social or cultural phenomenon, it is associated with the capacity of human societies to organise themselves through the formation of cities, states and other collective artefacts.

In recent years there has been a growing interest in studying and understanding phenomena like emergence, self-organisation and complexity. This interest is associated with the growth of complexity science. Complexity science strives to identify whether there are certain common principles that govern how components as diverse as atoms, cells, animals or humans 'organise themselves' and in so doing lead to the formation of macroscopic phenomena like chemical patterns (e.g. Prigogine and Stengers 1984), living structures (e.g. Kauffman 1969; Eigen and Schuster 1979), cognitive functions (e.g. Gazzaniga 1989; Kelso 1995), social (Bonabeau *et al.* 1999; Eberhart *et al.* 2001) or economic constructs (Anderson *et al.* 1988; Arthur *et al.* 1997). In short, complexity science aims to develop a set of methods and theories for understanding the organisational principles that underlie the creation of higher-level functions or structures (see for example Haken 1983; Badii and Politi 1997; Schuster 2001; Boccara 2004). The impact of complexity research in design studies has been quite sparse. Yet complexity science presents us with an interesting 'hypothesis' about the nature of design: that design can be perceived and studied as a macro effect of organisational dynamics. To date, there has been very little attention to this issue. This chapter aims to explore how complexity – as a theoretical and epistemological approach – may contribute to our understanding of the capacity to carry out design tasks in relation to the existence of certain organisational principles.

Certainly the development of a 'complexity' theory of design 'capacity' is not an easy feat. Currently there are only tenuous links between issues typically investigated in complexity research, such as stability, bifurcations, coordination, or phase transition, and the very notion of design. Within complexity science itself, the 'problem' or 'phenomenon' of design has been largely ignored. In this sense, it is necessary to attain a conceptual and mathematical framework that associates complexity theoretic constructions with issues pertinent to design. The focus of this study is to develop such a conceptual and mathematical construction.

This chapter is organised as follows. Section 2 discusses the scope of design phenomena and in particular elaborates the notion of design Intentionality in relation to ideas coming from philosophy of mind. Section 3 proposes a semantic theory of organisational complexity and then reintroduces the notion of design Intentionality as part of the proposed semantic theory of organisational complexity. In the final part (section 4), the mathematical conditions that underlie design Intentionality are explicitly identified.

2 Design Intentionality

The premise behind this treatment is that design arises in response to a certain 'problematic situation': when there is a desire, need or idea to construct a change in a certain environment, but the precise *means* and *ends* of this construction are not given. Although this premise is commonly held in design research (see for instance Archer 1965; Mitchell 1990; Smithers 2002), it can nevertheless take a number of different interpretations.

In the following, the premise that design arises in response to a particular situation is approached in relation to the capacity of an 'organism' to have 'Intentionality' or 'Intentional states'. Although the term 'Intentionality' has its origins in the writings of medieval scholastic philosophers, the contemporary 'technical' meaning of the term is reintroduced in philosophy of mind by Brentano in 1874 (republished in 1995). According to Brentano, Intentionality aims to describe the very essence of the mind: the capacity to represent or reflect objects or states of affairs in the world – either existing or non-existing. The capacity of a mind to represent an object or state of affairs in the world is 'Intentional' in the sense that mental states are 'semantically evaluable' and therefore 'refer to something', or they are 'about something' (Fodor 1995).

In philosophy of mind it is generally assumed that there are two archetypical, logically distinct 'Intentional states': *beliefs* about 'what the properties of a certain reality are', and *desires* about 'what the properties of a certain reality ought to be'. Although it is analytically convenient to distinguish different types of 'Intentional states', 'Intentionality' is generally perceived as a holistic or emergent property of the mind that is derived by a network of causally related Intentional states – i.e. a network of beliefs, desires or intentions. Based on these terms, and in accordance with our premise about the nature of design, a design situation can be seen as an Intentional state that 'emerges' out of *inconsistencies* between beliefs held about the past, current and future states of the world, and desires regarding the state of the world. However, in order to explicate this general intuition, it is important to introduce a common theoretical background on Intentionality.

The discussion takes as its starting point Brentano's core thesis that Intentionality is the mark of the mind. According to his position, Intentionality is the necessary and sufficient condition of mental states and therefore it is what distinguishes mental from physical phenomena. In contemporary philosophy of mind this claim is challenged in two different ways, leading to ontological questions such as: Do Intentional states correspond to physical processes or events? Are these physical processes or events in direct correspondence with the neuro-chemical processes and events occurring in the brain? Clarifying these objections and other core issues about the ontology of Intentionality is an opportunity to establish a common understanding on the proposed Intentional interpretation of 'design situations'.

2.1 The scope and structure of Intentionality

A first objection to Brentano's thesis is that mental states are not always Intentional. There are mental states that are not 'semantically evaluable', and therefore they are not directed to an object or state of affairs in the world. Typical examples are pain, general anxiety or elation. Although these are mental states, they are not reflections of a specific object or state of affairs in the world. In contrast, beliefs, desires, hopes or intentions are 'Intentional states' in the sense that they refer to something; the belief that 'it is raining' or the desire 'to stay dry' reflects a specific state of affairs in the world. In this description of Intentionality, there is a distinction between a 'mental state' and the objects at which it is directed or is about (Searle 1983). More specifically, the Intentionality of an organism is determined by two components: the *representational content* or *Intentional object* that describes the properties of an object or state of affairs in the world (e.g. 'it is raining' or 'stay dry'), and the *attitude* or *psychological mode* that determines the type of Intentional state (e.g. belief, desire or intention). This structure is often denoted as ***F***(s); where ***F*** is the attitude and s is the representational content. Alternatively, Fodor (1975, 1987) defines the attitude of an Intentional state as a binary relation ***F*** between an organism O and its mental states/mental representations s; that is ***F***(O,s). In this notation, s is a token, a syntactical physical entity, associated with an Intentional object (e.g. 'it is raining').

According to Searle (1983), an attitude ***F*** (e.g. beliefs or desires) expresses the underlying assumptions regarding the relative independence of the world from the mind. These assumptions determine a *direction of fit*. More specifically, an attitude ***F*** expresses 'a direction of fit' in the sense that any mismatch between the representational content (e.g. 'it is raining' or 'stay dry') and the world is resolved by either changing the mind in relation to the world, or changing the world in relation to the mind. Beliefs and desires are thus two archetypical Intentional states that express two opposite 'directions of fit'. A *mind-to-world* direction of fit assumes that an Intentional object represents an independently existing world, and therefore any mismatch between representation and the world must be followed by an adaptation of the representational content. A belief is either true or false in the sense that the correspondence between representational content and the world is evaluated against an independently existing world. A *world-to-mind* direction of fit assumes that an intentional object exists independently from the world, and any mismatch between the representational content and the world must be followed by an adaptation of the world. A desire cannot be said to be true or false. A desire is satisfied or fulfilled in the sense that the correspondence between representational content and the world is realised when there is a change in the world in relation to the mind. In both cases, the representational content s of an intentional state determines the *conditions of satisfaction* of a certain Intentional state. Based on these terms, Intentional states are therefore defined as mental states that have certain conditions of satisfaction s with a certain 'direction of fit' ***F*** in relation to an object or state of affairs in the world.

2.2 The semantic content of Intentionality

The second objection to Brentano's core thesis is that, although Intentionality is indeed a property of the mind, it is also a property of non-mental or physical objects. For instance, drawings, artefacts, sentences, or more generally expressions in a language, are 'semantically evaluable' (they refer to an object or state of affairs in the world) and therefore they are Intentional constructions. But drawings, artefacts or sentences are not mental states. The confusion is even greater when natural objects are perceived as 'natural signs' that hold information about other events (e.g. Millikan 2004). For instance, clouds can be perceived as natural signs that represent the possibility of rain. In this sense, both physical and mental states are distinguished for their semantic capacities and therefore for their Intentionality.

As a first response to these problems, a distinction is often made between 'intrinsic' and 'derived' Intentionality (e.g. Searle 1983). Mental phenomena are distinguished for their 'intrinsic Intentionality', while physical phenomena (such as language) are distinguished for their 'derived Intentionality'. Derived Intentionality is a type of Intentionality that is imposed on a physical object by the 'Intentional' states of a mind. According to this view, the formation of semantic relations is part of a theory of mind and a special form of Intentionality (Searle 1983; Fodor 1987). For Fodor (1987), Intentionality is ultimately a linguistic problem which may involve an 'inner' language – 'a language of thought' – but also any other 'outer' language. He starts with the assumption that mental states (i.e. the physical structure of the brain and its representational content) are expressed by a language and a symbol processing system. From this perspective, the distinction between the Intentionality of the brain and the Intentionality of any symbol system is reduced to the same problem. As he put it (Fodor 1987: vi): 'It would be therefore no great surprise if the theory of mind and theory of symbols were some day to converge'.

For Searle (1983: 160–79), the formation of semantics involves two layers of Intentionality, one that corresponds to a 'sincerity condition' and one to 'meaning intentions'. First, there is an Intentional state to be expressed by an object (e.g. the desire to stay dry may be expressed by a drawing or sentence); and, second, there is an 'intention' to express that Intentional state. Meaning is therefore acquired because of the intention of the mind to 'associate' an Intentional state with a non-mental or physical expression. This association is realised because the intention to express an Intentional state holds the same conditions of satisfaction with the expressed Intentional state.

In all these approaches, Intentionality is inextricably related to the formation of semantics. It is therefore important to delineate the semantic aspects of Intentionality and in particular how Intentionality leads to the formation of semantic relations.

2.2.1 Formal semantics

It is generally agreed that semantics alludes to the capacity of an expression (linguistic or mental) to hold and manipulate meaning. This is a view that is commonly held in philosophy of mind, linguistics and logic. The formal specification of semantic relations has been part of mathematical logic and in particular of 'model theory' (Barwise 1977). Model theory is concerned with the relation between a set of logical statements expressed in a language L and the mathematical structures that satisfy the postulated statements. The set of logical statements form a *theory*, and the algebraic structures that satisfy the statements of the theory are called *models*. A theory is consistent if it has at least one model. Assuming the existence of a model, a theory is a set of 'truth conditions' for that model. The distinction between *theories* and *models* is therefore the model theoretic way to explicate the formation of semantic relations as the interplay between 'the specification of the properties of a family of objects' and 'the set of objects that satisfy the properties of a specification'.

An alternative way to explicate the formation of semantic relations can be found in category theory (e.g. Barr and Wells 1985, 1990). Instead of employing a formal language in order to specify semantics (i.e. the truth conditions of a set of objects), category theory uses diagrams of objects, arrows and compositions of arrows. These diagrams are formally named *sketches*. In this chapter, category theory will be used in order to offer formal definitions of design, and so a more detailed introduction to the notion of sketch is necessary.

A sketch s is a graph G with certain additional structure. More specifically, this additional structure is expressed by imposing constraints in the commutativity of different paths of arrows in the graph G and by specifying levels of abstractions in the structure of the graph G. A sketch generates a mathematical category in the same way that recursive rules or a formal grammar generate a language (i.e. a set of expressions that satisfy certain grammatical rules). Based on this analogy, a category is an algebraic way to express the model theoretic notion of a theory. Similarly, a functor $\boldsymbol{F}$:A→B is the category theoretic way to explicate the model theoretic notion of a model. The functor $\boldsymbol{F}$ specifies a structure in B that preserves the properties of a theory A (or a sketch of a theory). The idea that categories can be used to explicate mathematical theories originates in the work of Lawvere (1963). The idea to use graph structures (or sketches) as a presentation of mathematical theories originates in the work of Ehresmann (1968).

More specifically, a sketch s = <G_S, D_S, L_S, C_S> is a graph G_S with a set of diagrams (graph homomorphisms) D_S that determine constraints on the composition of arrows over G_S; and a set L_S of cones and a set C_S of co-cones in G_S that generate a family of arrows with common properties. For each sketch determined by <G_S, D_S, L_S, C_S> it is possible to construct a category that has as an underlying graph the graph G_S; as commutative diagrams the set D_S; and as limits and co-limits the type of cones and co-cones defined in L_S and C_S. This category, denoted as ***Th***(s) or ***Th***s, is defined as a *theory of a sketch* s. For a sketch s and for a theory of a sketch ***Th***s there is a functor *i*:***Th***s→C that preserves the aforementioned properties of the theory. This functor is defined as a *model or interpretation of a theory* ***Th***s. In the same way, a *model of a sketch* s is defined as a sketch morphism *m*:s→***U***d from a sketch s to the underlying sketch ***U***d of the category d. A 'sketch morphism' is a graph homomorphism that takes the set of diagrams D_S, cones L_S and co-cones C_S to a set of diagrams, cones and co-cones in ***U***d.

Based on this notation, the following universal property is defined: for every sketch s and every model of a sketch *m*:s→***U***d there is unique functor *i*:***Th***s→d for which the following diagram commutes (i.e. two alternative paths of arrows are equal so, $m = \boldsymbol{U(i)}\ \eta_s$):

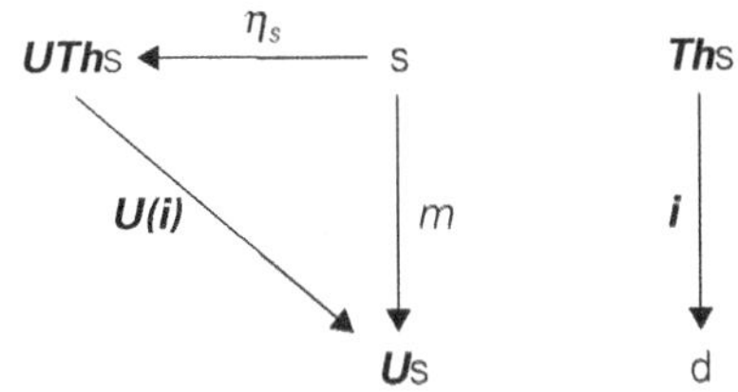

(1)

This property states that there is a natural bijection between models of a sketch m:s→***U***d and interpretations of a theory i:***Th***s→d (i.e. an adjunction between a theory functor ***Th*** and the underlying functor ***U***), which is denoted as:

$$\text{Mod}(s, \boldsymbol{U}\text{Cd}) \cong \text{Int}(\boldsymbol{Th}\text{s}, d) \qquad (2)$$

These category theoretic concepts can be used to explicate how an Intentional state leads to the formation of semantic relations. In particular, the basic idea is summarised in two levels:

First, there is an Intentional state that is intended to be expressed in a certain language D. This Intentional state is explicated by the model ***M***:s→***U***d and has certain conditions of satisfaction s. This state essentially corresponds to the first layer of Searle's Intentionality (Searle 1983: 160–79).

Second, there is an Intentional state (in particular an *intention*) that is realised by the functor ***F***:C→D between two categories C and D. C corresponds to the subjective reality and D to the objective reality. The functor ***F*** explicates the intention to express something in a language D given certain conditions of satisfaction expressed by the sketch s in C. This essentially corresponds to the second layer of Searle's Intentionality (Searle 1983: 160–79). A representation of the semantic properties of Intentional states is given in Figure 6.1.

To give an informal example of semantic relations, consider the 'five points for a new architecture' that Le Corbusier (1985) developed during the 1920s: (1) the pilotis; (2) the open plan; (3) the free façade; (4) the horizontal window; and (5) the roof garden (Figure 6.2).

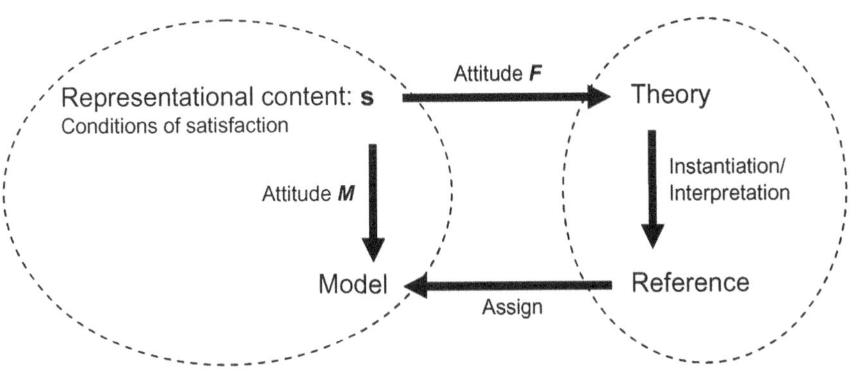

Figure 6.1
A schematic representation of the semantic properties of Intentional states. The category theoretic notion of sketch s is the representational content (or condition of satisfaction) of an Intentional state. The category generated by a sketch is the inner or outer language (a theory) used in order to express a specific Intentional state M

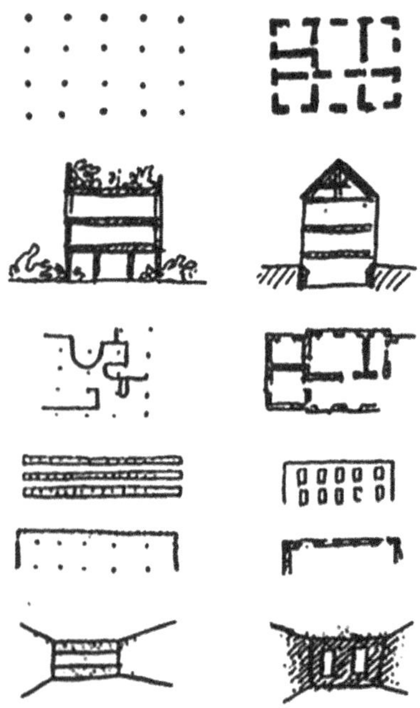

Figure 6.2 Le Corbusier's sketches illustrating his five points for a new architecture.
Source: http://www.en.sbi.dk/arkitektur/beredygtighed/arkitektur-og-beredygtighed-artikelsamling/de-signed-ecology

These five points (expressed in his sketches and written work) constitute the conditions of satisfaction (i.e. a sketch in a mathematical sense) that describes the properties of a desired reality. Based on these conditions, a theory of desired artefacts (in this case, buildings) is produced. For example, the sketches in Figure 6.3 illustrate a family of buildings that satisfy the five principles, thus exemplifying Le Corbusier's modernist language.

Each specific object that belongs to this family of buildings is described by a model. For example, the Villa Savoye, which is considered as an exemplary instantiation of Le Corbusier's theory, is in our terminology an object. Any description of the villa itself (in technical drawings, text or natural language) is a model of this object.

2.3 Realisations of Intentionality

Up to this point, the term Intentionality has been discussed as a special mental capacity; a capacity that is inextricably related to the formation of semantic relations. But what is the locus, body, or universe of Intentionality? How is Intentionality realised? Is the brain the only physical realisation of an Intentional mind? Do social entities have Intentionality? In order to respond to such questions, the 'problem' of Intentionality must be placed in relation to the 'mind–body problem'.

The mind–body problem concerns the relationship between mental states or processes and physical events. In contemporary philosophy and science, it is

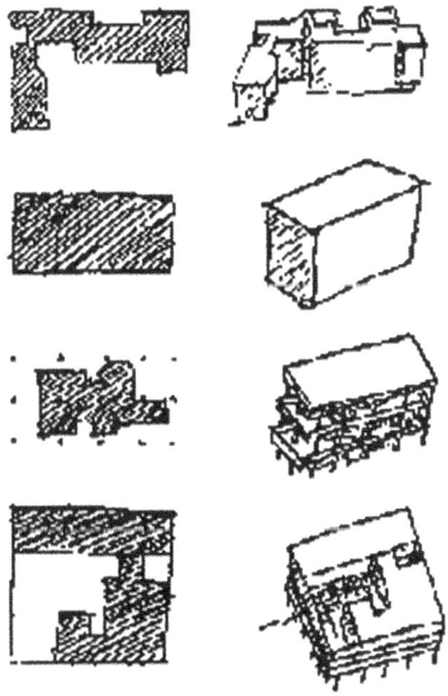

Figure 6.3 Exploration of the potentials of the five points by Le Corbusier: a family of buildings that exemplify the five points. Source: http://www.geocities.com/rr17bb/LeCorbusier5.html

commonly held that mental phenomena are somehow linked to physical phenomena, although the exact nature of this relation is not clear. There are four main approaches regarding this relation (e.g. Harman 1989; Chalmers 2002).

One approach, *behaviourism*, associates mental states and processes with the 'behavioural dispositions' or behavioural tendencies generated by the underlying physical processes of an organism. According to this view, the mind is a special aspect of the behaviour of a physical system (e.g. Ryle 1949; Dennett 1987).

The second approach, *identity theory*, associates mental states and processes with physical states and processes. More specifically, identity theory claims that mental states are essentially physical states of the brain of an organism (e.g. Place 1956; Smart 1959).

The third approach, *functionalist theory*, attributes mental states or processes to physical states or processes whose behavioural dispositions are distinguished for their functional role. A predominant interpretation of this approach postulates that mental states are functions of a computational machine (e.g. Putnam 1973). The view of mental states as functional entities of a computational machine helped explain how the same mental states might take multiple physical realisations. So, for instance, animals and humans may both have a similar Intentional state, but different neurological structures.

Finally, a number of alternative studies on the relation between mind and body have brought to the fore the *emergentist* hypothesis that mental states are higher-order/emergent properties of lower-level (typically physical) states and processes (e.g., for a review, see McLaughlin 1992, 1997; Horgan 1993). According to this approach, Intentional states are emergent qualities that cannot be deduced from the principles of the components found at a lower level of abstraction. This relation (between higher-order emergent properties and lower-level states and processes) is specified in different ways: as ontological, logical or epistemological relation (see Alexiou 2007). The emergentist view of the mind has been closely linked with the concept of 'supervenience', first introduced in the context of philosophy of mind by Davidson (1970). According to this view, the mind supervenes the physical states or processes of a brain: i.e. although the same mental states may have a different physical realisation, identical physical states or processes specify identical mental phenomena.

This short review shows that there is a large body of studies that support the conventional wisdom that the mind, and Intentional states in particular, can have many different realisations (e.g. Kim 1992). Intentionality can be manifested in different physical mediums, such as sounds or drawings (although one can always argue that this is a derived Intentionality as discussed above). Human or animal activities produce artefacts that are also Intentional. Artefacts are 'Intentional constructions' in the sense that they are physical realisations of the beliefs and desires of an Intentional mind. However, artefacts are not always the product of an individual organism. Social constructions such as ant nests, human cities or nations are artefacts that express the beliefs and desires of a society. These artefacts are obviously the products of some Intentional mind, although this Intentionality cannot be attributed to a specific brain physiology.

The possibility of collectively attributing Intentionality to a physical entity entails an Intentional state that cannot be broken down to individual Intentional states. (In the same sense that Individual Intentionality is often studied as a holistic state that cannot be broken down to individual attitudes – beliefs, desires, intentions etc.) It is a rather new form of Intentionality that arises as the aggregate effect of Intentional organisms. In this sense, Intentionality can be said to be embodied in certain organisational structures and processes that are not necessarily realised in the physical structure of a brain. For the purpose of this study, the term Intentionality or Intentional state will be used for any 'organism' whose physical or social realisation has the capacity to hold representations of an 'objective' reality. In this sense, this distinction between 'representation' and the 'representing object' marks the distinction between a subjective and an objective reality. The ontological status of these realities is not specified, but the logical properties of this coupling will be the main focus of the next sections.

3 An organisational-level description of 'design Intentionality'

We can now move on to consider how an organisational-level description of design Intentionality can be attained. Before we are able to do this, it is necessary to introduce a semantic/Intentional theory of organisational complexity: a mathematical framework for expressing semantic aspects of complexity. The second part of this section is an application of the proposed theory in the description of design Intentionality.

3.1 A short introduction to a semantic theory of organisational complexity

As mentioned in the introduction, the fundamental objective of complexity science is to identify whether there are *universal principles* that govern how components as diverse as atoms, cells, animals, or humans 'organise themselves' and in so doing lead to the formation of macroscopic phenomena like chemical patterns, living structures, cognitive functions or social constructs.

There are many different approaches to developing a mathematical description of organisational complexity. One approach is to focus on *ontological properties* of organisation – such as, for instance, the degrees of freedom and mutual information in the description of a system (e.g. Nicolis and Prigogine 1967; Atlan 1974; Von Foerster 1984). Another approach is to focus on the *logical or computational properties* of organisation – such as, for instance, the computational resources (the time or logical effort) needed in order to describe/compute a system (e.g. Kolmogorov 1965; Chaitin 1966, 1997; Bennett 1985), or the computability of a system (e.g. Langton 1990; Wolfram 1994). A third approach is to focus on the *epistemological aspects* of organisation – such as, for instance, the characterisation of the number of independent descriptions or independent models of a system (e.g. Rosen 1991; Cariani 1991; Casti 1992).

For the purpose of this study, the focus is placed on the *semantic aspects* of organisation. The proposition put forward is that the notion of the sketch presented above can be used in order to describe the organisational complexity of a system. More precisely, the universal property of semantic relations (that is the adjunction that arises between syntax and semantics) is employed as a tool for the characterisation of organisational complexity. According to this perspective, an *'ordered' organisation* is an organisation that is described by a well-formed theory; a *'random' organisation* is an organisation that is characterised by the absence of such a theory; and a *'complex' organisation* is an organisation that is described by a notion of a weak theory. A weak theory can then be thought of as the most general class of theories which allows us to consider well-formed and random theories as special cases.

For the purpose of defining the notion of weak theory, it is important to specify the notion of weak adjunction. More specifically, the concept of weak adjunction between two categories C and D requires the existence of two functors ***F*** and ***U*** with the following properties:

Definition (weak adjunction): A weak adjunction between two categories C and D is defined by a tuple $<\boldsymbol{F}, \boldsymbol{U}, \varphi, \theta, \tau_C, \tau_D>$ where $\boldsymbol{F}$:C→D and $\boldsymbol{U}$:D→C are functors; the arrow τ_C is a (natural) transformation between the arrows (functors) m:s→**U**d and m':s→**U**d; the arrow τ_D is a (natural) transformation between the arrows (functors) i:**F**s→d and i':**F**s→d; and the arrows φ and θ make the following diagram commute naturally in s and d (i.e. $\varphi°\theta = \tau_C$ and $\theta°\varphi = \tau_D$):

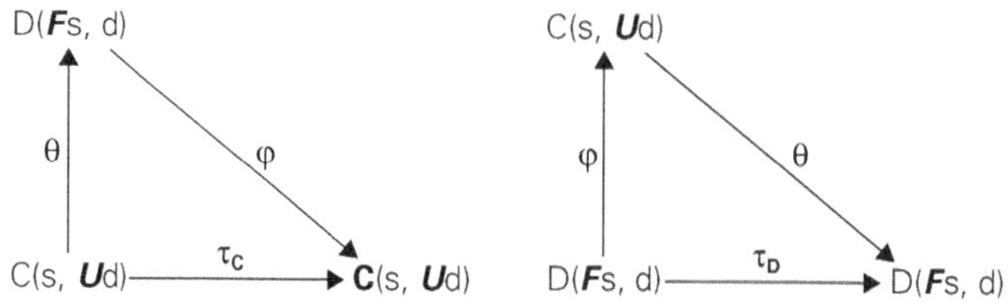

(3)

The two diagrams can be equally depicted as follows:

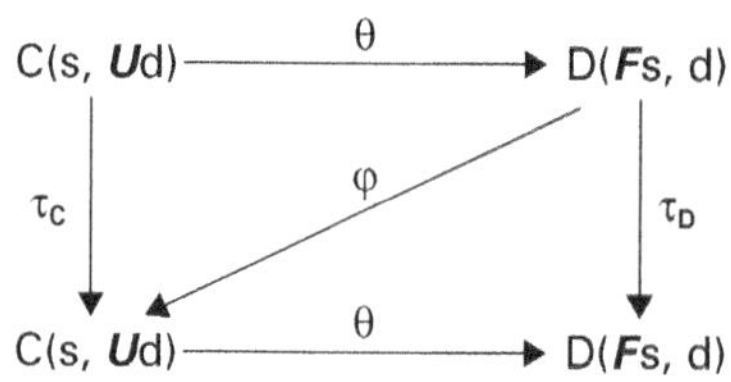

(4)

As usual, the condition of naturality for arrow φ (and similarly for θ) means that the following diagrams also commute for every arrow m:s′→s in C and i:d→d′ in D:

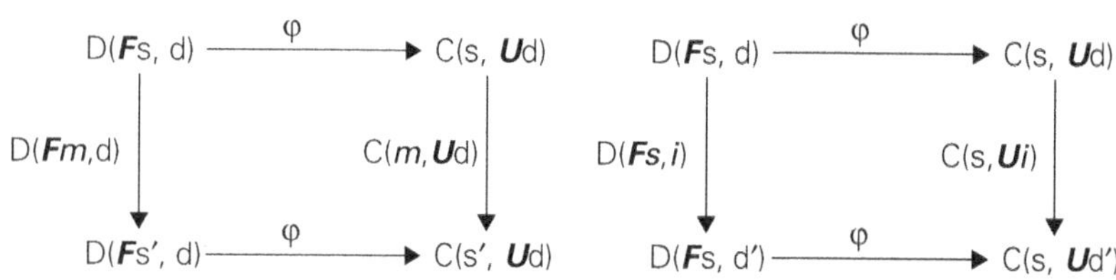

(5)

Based on this definition, the following special cases can be defined:

- If there is an object s in C or d in D with $\tau_C = 1_{C(s,Ud)}$ or $\tau_D = 1_{D(Fs,d)}$ – where $1_{C(s,Ud)}$ and $1_{D(Fs,d)}$ are identity arrows over C(s, **U**d) and D(**F**s,d) respectively – then for the objects s in C and d in D there is a universal arrow η_S:s→**UF**s and ε_d:**FU**d→d respectively.
- If for every object s in C or d in D the arrows $\tau_C = 1_{C(s,\ Ud)}$ or $\tau_D = 1_{D(Fs,\ d)}$ – that is, when the arrows τ_C and τ_D are identity arrows for every object in C or D – then θ and φ form a bijection that is natural in s and d (hence the tuple <**F**,**U**,φ,θ> is an adjunction).

Based on this construction, a weak theory is defined as follows:

Definition (weak theory): A weak theory is a category **Th**s that is constructed by a sketch s in C, and a functor **Th** such that a weak adjunction <**Th**,**U**,φ,θ,τ_C,τ_D> is defined; i.e. the relation φ and θ between interpretations i of theories **Th**s in D and models m of a sketch s in C are determined by the following diagram:

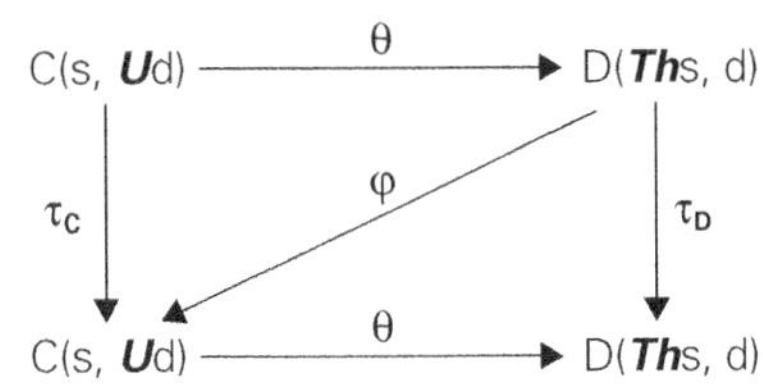

(6)

Well-formed and *random* theories can be thought of as special cases of weak theories in the following sense:

- A well-formed theory is constructed when for every object s in C or d in D the arrows $\tau_C = 1_{C(s,\, Ud)}$ and $\tau_D = 1_{D(Ths,d)}$ – that is, when the arrows τ_C and τ_D are identity arrows for every object in C or D. In this case the arrows θ and φ form a bijection that is natural in s and d, and the tuple $<\boldsymbol{Th}, \boldsymbol{U}, \varphi, \theta>$ is an adjunction.
- A random theory is constructed when there is no object s in C or d in D such that $\tau_C = 1_{C(s,\, Ud)}$ or $\tau_D = 1_{D(Ths,d)}$. In this case the adjunction $<\boldsymbol{Th}, \boldsymbol{U}, \varphi, \theta>$ is broken.

The notion of weak theory can now be used to build a mathematical construction that describes a phase transition in the behaviour of a (mathematical) universe U. A phase transition is perceived as a transformation of the properties and degree of complementarity between descriptions (theories) and their interpretations (models).

Definition (phase transition): given the arrows T_s:s→s′ and T_d:d→d′ shown below,

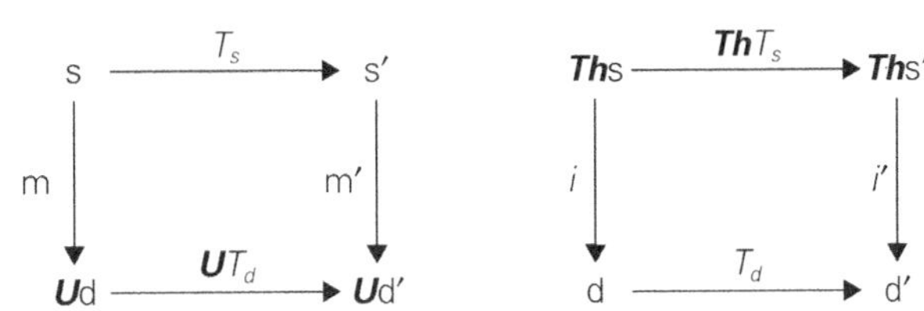

(7)

phase transition is defined by the transformations $T_C = C(T_s, \boldsymbol{U}T_d)$:C(s,**U**d)→C(s′,**U**d′) and $T_D = D(\boldsymbol{Th}T_s, T_d)$:D(**Th**s,d)→D(**Th**s′,d′) that make the following diagram commute:

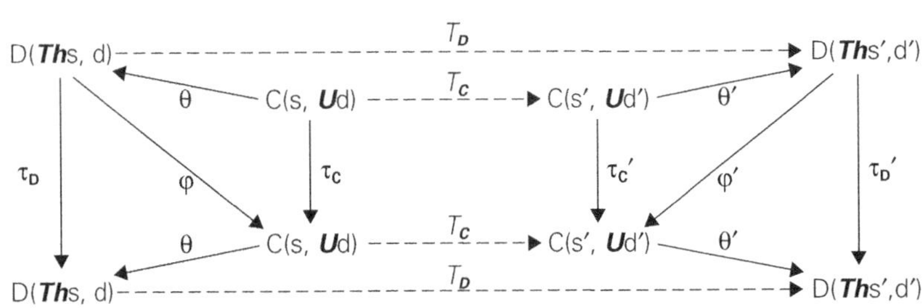

(8)

3.2 Organisational-level description of design Intentionality

The concepts of weak theory and model can now be used in order to explicate the view of design as an Intentional state of a mind at an organisational level of description.

First, let us redefine the notion of 'design situation' outlined in section 2. *A problematic (or design) situation arises when certain desires about the world generate expressions of theories or models (of objects) that do not follow the correspondence between theories and models as this is established by one's belief system.* In this context, a sketch S is the conditions of satisfaction of a subjective reality (i.e. of a mind) that produces an – often incomplete, inconsistent or unsatisfied – expression of an objective reality.

However, a sketch of a design situation S_D is an Intentional state of a particular kind. It is an Intentional state with a 'two-directional freedom of fit' (recall Searle's notion of 'direction of fit'). On the one hand, a sketch of a design situation expresses the conditions of satisfaction of a desire that is *fulfilled* only when there is a theory of an objective reality that satisfies its models. On the other hand, a sketch S_D of a design situation expresses the conditions of satisfaction of a belief that becomes *true (or false)* only when there is a model that validates the theory. Namely, a sketch S_D of a design situation is a particular type of Intentional state that arises when there is mismatch between a subjective and objective reality, and both the subjective and objective reality need to be adapted. So a ***problematic or design situation*** *arises when there is an underlying sketch* S_D *that generates a mismatch between an objective and subjective reality that can only be satisfied (i.e. fulfilled and validated) with the formation of both new theories as well as new models.* To put it differently, a design situation arises when the diagram in Figure 6.4 *does not commute.*

More formally, given a sketch s, the theory functor ***Th*** determines a category ***Th***s that constitutes a theory of plausible or desirable interpretations of a sketch. The functor ***U*** determines a category ***U***d that constitutes the set of properties of a specific interpretation of a theory ***Th***s in C. Models m:s→***U***d are those interpretations of the sketch that satisfy these properties. Hence, in category theoretic terms, a problematic situation appears when the developed theory ***Th***s of desired objects/processes yields interpretations in C (i.e. i:***Th***s→d) whose underlying properties ***U***d cannot be

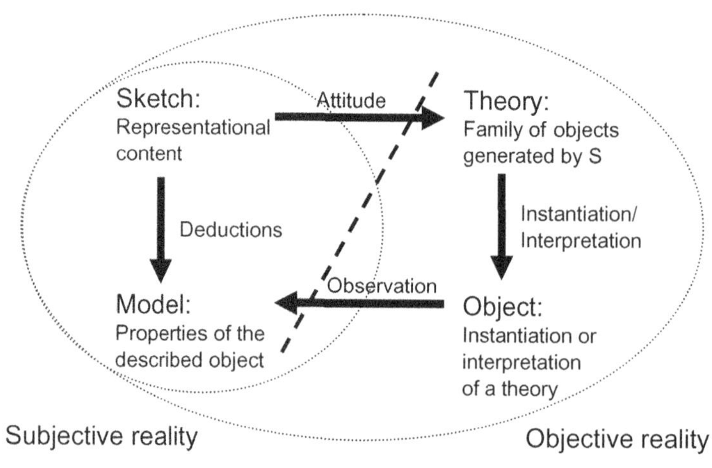

Figure 6.4 A sketch as an 'Intentional state'. A design situation arises when there is a 'two-directional freedom of fit' requiring the adaptation of both subjective and objective reality

(uniquely) derived by the sketch s (i.e. m:s→***U***d). The need to design therefore arises when the theory and model functors contain – so to speak – ambiguity, noise or errors, and hence there is no natural bijection between C(s,***U***d) and D(***Th***s,d). The following diagram is an informal illustration of this situation.

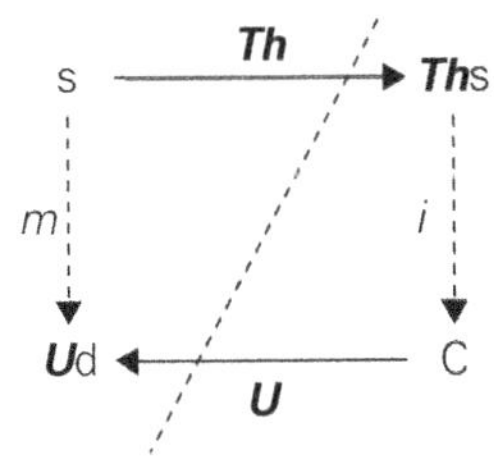

(9)

More precisely, a design situation is defined when there is weak adjunction (no unique correspondence):

$$\mathrm{Mod}(s,\boldsymbol{U}d)\not\cong\mathrm{Int}(\boldsymbol{Th}s,d) \tag{10}$$

In response to a problematic situation, the 'task' of design is identified with the generation of theories and models of sketches that bring beliefs and desires into correspondence. ***The task of design*** *implies the existence of a sketch s that marks the transition from models with incomplete, inconsistent or unfulfilled theories, to models with complete, consistent and satisfactory theories.* In this sense, the need to design collapses when there is a sketch that matches the subjective with the objective reality in both directions: from world to mind and from mind to world. The definition expresses a generally accepted premise regarding the nature of design. For instance, according to Smithers (2002), at the core of design(-ing) lies an apparent paradox: designing has to do with arriving at a solution to a problem which is not *a priori* specified. In other words, although design is driven by a need or a goal, this goal is actually constructed by the very process of design.

What surfaces from this discussion is that the peculiarity of design lies in the fact that sketches, models and theories are developed in some way independently from each other. More precisely, the postulation of a weak adjunction implies that the construction of theories and the construction of models from a sketch also lead to certain expressions of desired objects/processes that are not uniquely derived from the specified sketch. In this sense, design necessitates the (paradoxical) capacity of generating theories and models of a sketch in preparation of their adjunction, i.e. before such correspondence is constructed. In other words, design requires the capacity to generate sketches *in anticipation of* the theory–model adjunction (Zamenopoulos and Alexiou 2007). The main results from this treatment can now be formally summarised.

3.2.1 Definition: design situation

A design situation is defined as an Intentional state s with a 'two-directional freedom of fit' – in a subjective C(_,***U***_) and objective reality D(***Th***_,_) – whose models (i.e. m:s→***U***d) in C(s,***U***_) and interpretations (i.e. i:***Th***s→d) in D(***Th***_,d) are not complementary. Namely, a design situation arises when there is a weak adjunction <***Th***,***U***,$\varphi,\theta,\tau_C,\tau_D$> natural in s and d that makes the following diagram commute:

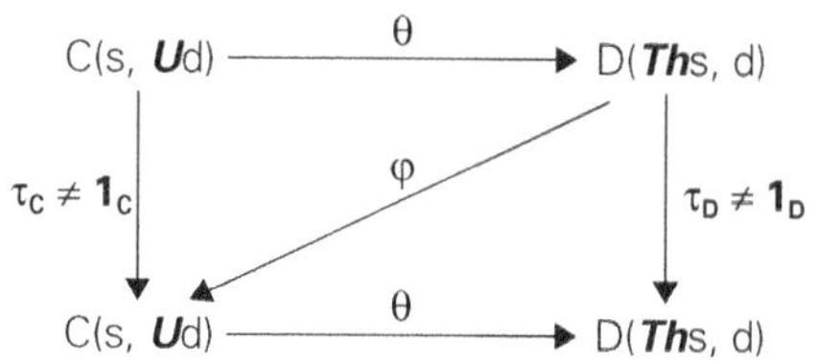

(11)

Given a design situation, the 'problem' of design or the 'design task' is to establish an adjunction Mod(s,***U***d)≅Int(***Th***s,d) by transforming both the subjective and objective reality.

3.2.2 Definition: the design task

The problem or task of design is identified with the existence of a phase transition to universality; that is the existence of transformations

$T_C = C(T_s, \mathbf{U}T_d)$:C(s,***U***d)→C($s_U$,***U***$d_U$) and $T_D = D(\mathbf{Th}T_s, T_d)$:D(***Th***s,d)→D(***Th***s_U,d_U)

such that the following diagrams commute:

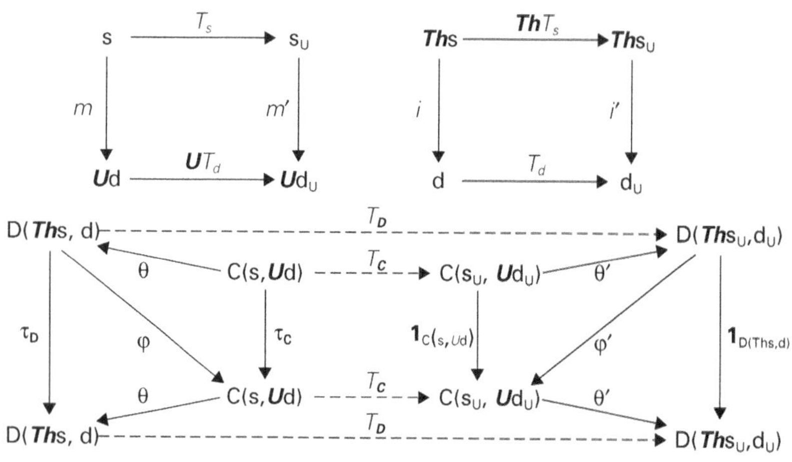

(12)

Based on this definition of a design task, it is plausible to assume that, given an 'Intentional state' s_D, the *capacity to carry out design tasks* implies the existence of transformations T_S:s_D→s_u in a subjective reality C(_,***U***_) and transformations T_d:d_D→d_U in an objective reality D(***Th***_,_) that are *components of a phase transition to universality* $T = <T_C, T_D>$. The transformation T_S:s_D→s_u is a component of the transformation $C(T_s, \mathbf{U}T_d)$ = T_C:C(s,***U***d)→C(s_U,***U***d_U), and similarly the transformation T_d:d_D→d_U is a component of the transformation $D(\mathbf{Th}T_s, T_d)$ = T_D:D(***Th***s,d)→D(***Th***s_U,d_U). It is informative to think that the transformation T_S:s_D→s_u has a 'mind-to-world' direction of fit, in the sense that any mismatch between a subjective C(_,***U***_) and objective reality D(***Th***_,_) is addressed by changing the properties of the sketch. Similarly, the transformation T_d:d_D→d_U has a 'world-to-mind' direction of fit because any mismatch between the subjective and objective reality is addressed by changing the properties of the object.

Let us now assume that the theory ***Th*s** generated by a sketch s specifies a hom-functor D(***Th*s**,d):C^{op}xD→**Set** that is naturally isomorphic with the functor C(s,***U*d**):C^{op}xD→**Set**; so C(s,***U*d**)≅D(***Th*s**,d). In category theoretic literature, the theory ***Th*s** is said to be a *representing object* for the hom-functor C(s,***U*d**):C^{op}xD→**Set**. Similarly, the underlying sketch ***U*d** of an entity d is said to be a *representing object* for the hom functor D(***Th*s**,d):D→**Set**. But what can be said for hom-functors D(***Th*s**,d) and C(s,***U*d**) that are not naturally isomorphic, that is when C(s,***U*d**)≇D(***Th*s**,d)? In this case, the notion of an anticipatory representation is defined. The theory ***Th***s_D generated by a sketch s_D is called an *anticipatory representation* of models in C(s_U,***U***d_U) *if and only if the sketch (Intentional state)* s_D *has models* m_D *and interpretations* i_D *that are components of a phase transition to universality* $T=<T_C,T_D>$. The underlying sketch ***U*d** is called an anticipatory representation of interpretations in D(***Th***s_U,d_U). Based on this definition, *the capacity to carry out design tasks* is now identified as the capacity to hold anticipatory representations/interpretations of universality.

3.2.3 Definition: the capacity to design

The capacity to design is the capacity of an Intentional state s_D to hold models T_C:s_D→s_U and generate interpretations ***ThT***$_C$:***Th***s_D→***Th***s_u that are components of a phase transition to universality (i.e. that satisfy the diagrams in definition 3.2.2). The capacity to design is therefore the capacity of a sketch s_D *to* generate theories ***Th***s_D of a subjective reality s_D and sketches ***U***d_D of an object d_D that are anticipatory representations of a universal construction C(s_U,***U***d_U)≅D(***Th***s_U,d_U).

The models m_D and theories ***Th***s_D that are derived from an Intentional state s_D are specified *in relation to a phase transition to universality.* Such anticipatory representations are clearly distinct from 'ordinary' representations where models and theories are specified *in relation to a universal construction.* More specifically, the meaning of anticipatory representation implies that a model m_D:s_D→***U***d_D and interpretation i_D:***Th***s_D→d_D are defined in 'preparation' of a universal C(s_U,***U***d_U)≅D(***Th***s_U,d_U); that is when the domain of both transformations T_c and T_D

$$\begin{array}{ccc} s & \xrightarrow{T_s} & s_U \\ {\scriptstyle m}\downarrow & & \downarrow{\scriptstyle m'} \\ \boldsymbol{U}d & \xrightarrow{\boldsymbol{U}T_d} & \boldsymbol{U}d_U \end{array} \qquad \begin{array}{ccc} \boldsymbol{Th}s & \xrightarrow{\boldsymbol{Th}T_s} & \boldsymbol{Th}s_U \\ {\scriptstyle i}\downarrow & & \downarrow{\scriptstyle i'} \\ d & \xrightarrow{T_d} & d_U \end{array} \qquad (13)$$

has co-domain or target domains that form a part on a universal construction C(s_U,***U***d_U)≅D(***Th***s_U,d_U).

The adjunction C(s_U,***U***d_U)≅D(***Th***s_U,d_U) uniquely characterises a design situation s_D in the sense that every model of s_D and interpretation of theories ***Th***s_D are constrained by the specified transition to universality. However, the reverse is not true. The models of the sketch s_D are not universal representations of $D(\boldsymbol{Th}s_U,d_U)$. It is in this sense that the adjunction $C(s_U,\boldsymbol{U}d_U)\cong D(\boldsymbol{Th}s_U,d_U)$ is perceived as an 'emergent' state of s_D.

4 Mathematical conditions of design Intentionality

The main thesis derived from the present enquiry is that the capacity to recognise and carry out design tasks implies the capacity of an 'Intentional state' (or sketch s_D) to construct anticipatory representations of a phase transition to universality. In this context, the capacity to construct representations of a phase transition to universality was explicitly associated with the capacity to form anticipatory representations. But what are the mathematical conditions that are responsible for the capacity of an Intentional state to hold anticipatory representations? What are the organisational conditions that are responsible for the capacity to carry out design tasks? This section aims to address these questions and, more specifically, to identify the mathematical structures that underlie the construction of an anticipatory representation of universality.

Anticipation is generally associated with the ability to look ahead (or look forward), but it also refers to an action or decision that is taken in preparation for some future event. Cambridge Dictionary Online (2006) defines anticipation as follows: 'to imagine or expect that something will happen, sometimes taking action in preparation for it happening' (for a review, see Zamenopoulos and Alexiou 2007). For the purpose of this section, anticipation is defined as the capacity – embedded in A – to represent an arrow $\boldsymbol{F}$:A→B (i.e. efficient cause) in preparation for certain effects in B (von Glasersfeld 1998).

$$A \xrightarrow{?F} B \tag{14}$$

But what does the expression 'in preparation for' mean precisely? And, moreover, how does this view of anticipation relate to the definition of anticipatory representations given at the beginning of this chapter?

Rosen (1991) has offered a mathematical interpretation of this view of anticipation and an explanatory framework of how this anticipatory capacity is possible. Rosen's key idea is that any expression s in A is a representation of a function φ_f:B→H(A,B); where H(A,B) denotes the set of all possible arrows $\boldsymbol{F}$:A→B. For any object d in B, the function φ_f:B→H(A,B) is then a representation of a functor from A to B that satisfies the bijection $A(s,\boldsymbol{U}d)\cong B(\boldsymbol{F}s,d)$. The dotted arrows in the following diagram are meant to depict this idea:

$$A \xrightarrow{\boldsymbol{F}} B \xrightarrow{\varphi f} H(A,B) \tag{15}$$

This view of anticipation can be perceived in relation to the definition presented in the previous sections. The arrow φ_f:B→H(A,B) effectively 'bounds' or drives the system in the bijection A(s,***U***d)≅B(***F***s,d). More specifically, the arrow φ_f:B→H(A,B) realises a transition to a universal state. Rosen observes that this representation is mathematically possible under certain conditions. The purpose of the next sections is to clarify these ideas and the underlying mathematical conditions. The precise mathematical justification of these conditions can be found in Letelier *et al.* (2006). So let us start with the general meaning of the proposed conditions.

4.1 An interpretation of Rosen's main results on anticipation

Rosen's main idea is that the capacity to anticipate is possible if and only if there is an embedding of the set A into the set of arrows φ_f:B→H(A,B). The set of arrows φ_f:B→H(A,B) is denoted H(B,H(A,B)) and the embedding by the arrow *e*:A→H(B,H(A,B)). The embedding *e*:A→H(B,H(A,B)) explicates the idea that the elements of the set A are anticipatory representations of the universal state A(s,***U***d)≅B(***F***s,d). The mathematical conditions that specify the embedding e:A→H(B,H(A,B)) are therefore the mathematical conditions that explain the construction of anticipatory representations.

In order to explain these conditions, note that the set of arrows H(B,H(A,B)) in the embedding *e*:A→H(B,H(A,B)) is the 'opposite' set of arrows from the set H(H(A,B),B). This letter set is an important set because it holds a special type of arrows; the evaluation arrows. Assuming that A and B are sets, an evaluation arrow ε_S is a function ε_S:H(A,B)→B that gives for every function ***F*** in H(A,B) the value of ***F*** at s in B, that is:

$$\begin{aligned} \varepsilon_S : H(A,B) &\longrightarrow B \\ \boldsymbol{F} &\mapsto \varepsilon_S(\boldsymbol{F}) = \boldsymbol{F} \circ s = d \end{aligned} \qquad (16)$$

More generally, the evaluation arrow is a universal arrow ε:***F***∘$d^{\boldsymbol{F}}$→d such that the following diagram commutes:

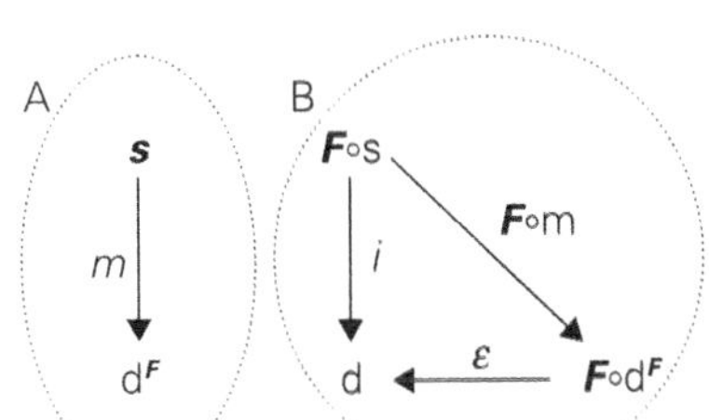

(17)

Every object s in A is therefore also a 'generator' or a 'program' that produces an object d in B for a given ***F*** in H(A,B). More importantly, there is a unique correspondence between objects s of A and the universal arrow ε_S:H(A,B)→B. If A and B are sets, then there is a one-to-one correspondence between the elements of the set A and the evaluation arrows ε_S:H(A,B)→B. Based on this observation, Rosen's main result is that the embedding *e*:A→H(B,H(A,B)) is defined when the evaluation function ε_S:H(A,B)→B in H(H(A,B),B) is invertible (one-to-one). In that way, the elements of the set A 'hold information' about the function F only by looking at the value of ***F*** at s. Based on this information, every element s of A determines a function F in preparation of a certain value d in B such that $A(s,d^F)\cong B(\mathbf{F}\circ s,d)$ (i.e such that $\mathbf{F}\circ s = d$ for $s = d^F$). This is the very meaning of anticipation. Rosen's condition can be given a weaker interpretation: the embedding *e*:A→H(B,H(A,B)) is defined when for each arrow $\mathbf{F}\circ s\to d$ there is an opposite universal arrow $\alpha:d\to\mathbf{F}\circ d^F$ that makes the following diagram commute.

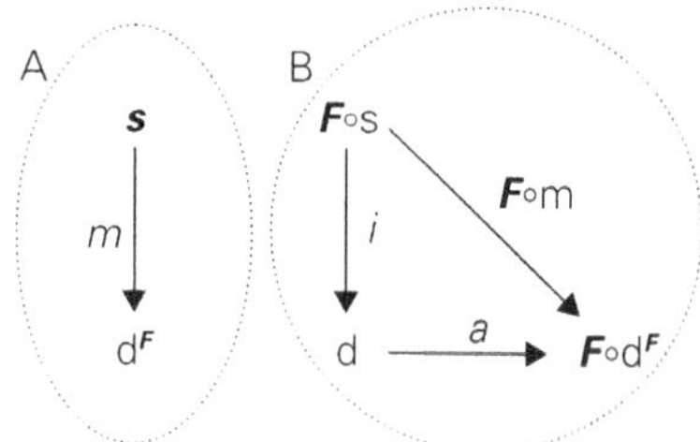

(18)

In this weaker version, the arrow $\alpha:d\to\mathbf{F}\circ d^F$ is not necessarily the inverse of the evaluation arrow $\varepsilon:\mathbf{F}\circ d^F\to d$. The postulated isomorphism $\mathbf{F}\circ d^F\cong d$ is therefore only a special case. Nevertheless, the arrow $\alpha:d\to\mathbf{F}\circ d^F$ holds information about the evaluation function $\varepsilon:\mathbf{F}\circ d^F\to d$ by specifying the relation between the arrows m, i and α as suggested by the above diagram (i.e. $\mathbf{F}\circ m = \alpha\circ i$).

4.2 Design as the capacity for 'world-to-mind' and 'mind-to-world' adaptation

At the beginning of this chapter, the capacity to design was identified with the capacity to hold anticipatory representations of universality with a two-directional degree of freedom, including 'world-to-mind' adaptations where there are transformations over an objective reality d_D to d_U, but also 'mind-to-world' changes where there are transformations from intentional state s_D to an intentional state s_U. If we go back to Rosen's original diagram (below), we can see that expressions s in A remain un-entailed: namely, changes of expressions in A are still to be dictated by the 'structure' of the environment E. It is clear that an additional closure condition needs to be formulated; this is the entailment of A.

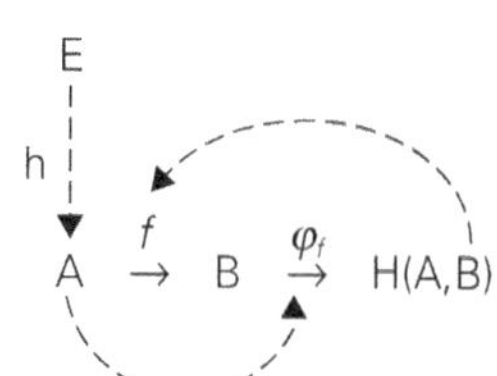

(19)

The proposition put forward here is to postulate the existence of two parallel structures realised by the embedding e_A:A→H(B,H(A,B)) and the embedding e_B:B→H(H(A,B),A) as the following diagrams suggest.

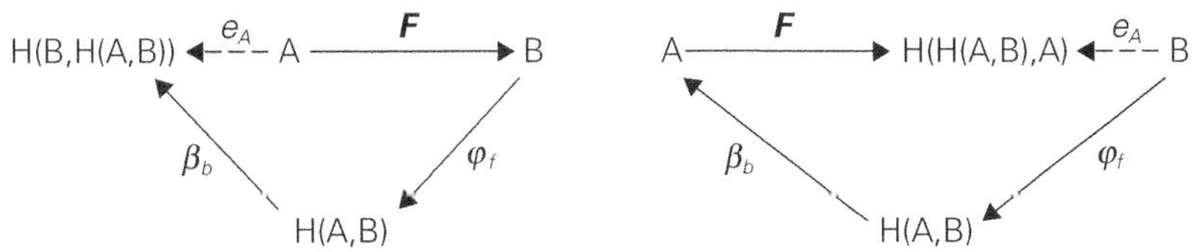

(20)

These conditions can be generally inferred and explained by supposing that Rosen's results about B are also applied to A. Given this assumption, any intentional state s_D in A represents an arrow φ_f in H(B,H(A,B)), and any object d in B represents an arrow β_b in H(H(A,B),A). The embedding e_A:A→H(B,H(A,B)) implies the capacity of an intentional state s in A to represent a functor ***Γ*** (and therefore specify an objective reality d in B) in preparation of certain intended properties in B.

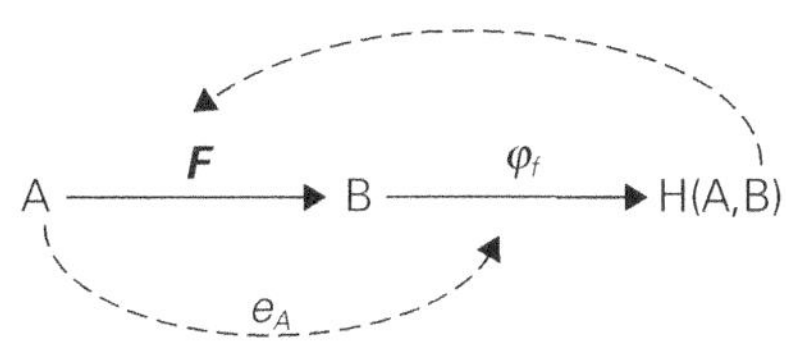

(21)

The embedding e_B:B→H(H(A,B),A) implies the capacity of an object d in B to represent a functor β_b (and therefore specify an Intentional state s in A) in preparation of certain properties in B (below).

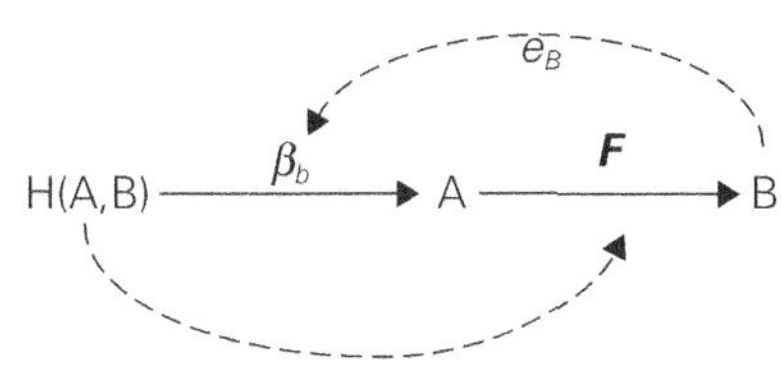

(22)

4.3 Summary

As a summary, consider that a functor ***F*** is in principle a (weak) theory functor ***Th***. Similarly, an object d^S in A represents the underlying properties of the object d in B. The capacity to design then refers to the capacity of an Intentional state s to hold models and interpretations that mark a transition to universal models η_s:s_U→***UTh***s_U. The core meaning of 'world-to-mind' adaptation refers to the existence of an Intentional state s that represents the transition from a weak theory ***Th***s_W in B to a functor ***Th***$_U$ in H(A,B) such that A(s_U,***U***d_U)≅B(***Th***s_U,d_U). The core meaning of 'mind-to-world' adaptation refers to the specification of an objective reality that represents a transition from a weak theory function ***Th***$_D$ in H(A,B) to an Intentional state s_U in A such that A(s_U,***U***d_U)≅B(***Th***s_U,d_U).

This 'universe' specified by an objective reality B and a subjective reality A has the capacity to represent and carry out design tasks if and only if the embeddings $e_A{:}A \rightarrow H(B,H(A,B))$ and $e_B{:}B \rightarrow H(H(A,B),A)$ exist. The embeddings $e_A{:}A \rightarrow H(B,H(A,B))$ and $e_B{:}B \rightarrow H(H(A,B),A)$ are then defined when for each Intentional state s and each arrow $i{:}\boldsymbol{Th}s \rightarrow d$ there is an (opposite) universal arrow $\alpha{:}d \rightarrow \boldsymbol{ThU}d$ such that the following diagram commutes:

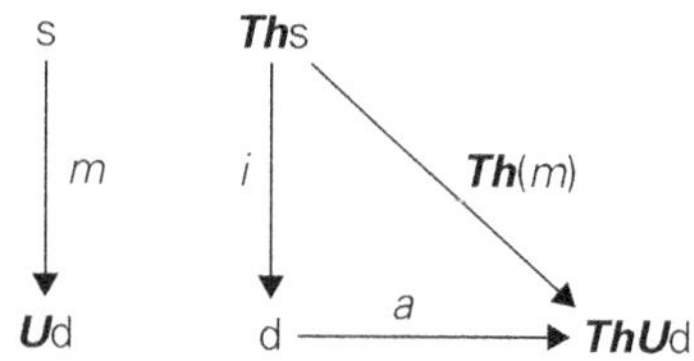

(23)

5 Conclusions

The main contribution of this chapter is the introduction of a mathematical theory of design that explains the capacity to carry out design tasks in terms of organisational complexity. This result is the product of three main developments:

- The chapter proposes a mathematical framework for understanding organisation and complexity. In particular it proposes a semantic theory of organisation that can encompass the Intentional character of design.
- The chapter explicates the notion of design Intentionality and links it to complexity research.
- The chapter identifies and defines the mathematical principles of organisation that are responsible for the capacity to recognise and carry out design tasks.

The contribution of this theory is that it proposes an understanding of design as a 'natural' phenomenon that is related with the complexity of organisation. It sets the conceptual foundations for studying design not only across different domains of human activity (e.g. art, science or engineering) but also across different levels of realisation (i.e. cognitive, social or cultural). Future investigations may focus on different biological, cognitive or social systems, and attempt to derive different types of organisational descriptions. For example, it is possible to use the category theoretic constructions developed in this chapter in order to study the organisation of brain activities that accompany design tasks (i.e. to study design at a neurological level), or to study the organisation of design functions and activities developed in social systems (i.e. to study design at a social level). Another route of future investigation is to use the theory in order to explore the possibility of creating artificial design-capable systems. Is it possible to construct artificial design systems that go beyond the computational approach? What (new?) elements can support such an endeavour (material, digital, analog, quantum)? Finally, the elaboration of design as a universal capacity of complexity is a double contribution to complexity research as well as to design research, and makes it possible to transfer results between the two fields. The proposed framework helps place the problem of design within the realm of complexity research and vice versa. This is of course a very large and ambitious research programme and, very importantly, one that can only be achieved with work that cuts across research domains and disciplines.

References

Alexiou, A. (2007), 'Understanding multi-agent design as coordination', in *Bartlett School of Graduate Studies*, London: University College London, University of London.

Anderson, P. W., Arrow, K. J. and Pines, D. (eds) (1988), *The Economy as a Complex Evolving System*, New York: Perseus Books.

Archer, B. L. (1965), 'Systematic methods for designers, in N. Cross (ed.), *Developments in Design Methodology*, Chichester: John Wiley, pp. 57–82.

Arthur, W. B., Lane, D. and Durlauf, S. (eds) (1997), *The Economy as a Complex Evolving System II*, New York: Perseus Books.

Atlan, H. (1974), 'On a formal definition of organization', *Journal of Theoretical Biology*, 45 (2): 295–304.

Badii, R. and Politi, A. (1997), *Complexity: Hierarchical Structures and Scaling in Physics*, Cambridge: Cambridge University Press.

Barr, M. and Wells, C. (1985), *Toposes, Triples and Theories*, New York: Springer.

Barr, M. and Wells, C. (1990), *Category Theory for Computing Science*, Englewood Cliffs, NJ: Prentice Hall.

Barwise, J. (1977), *A Handbook of Mathematical Logic*, Amsterdam: North-Holland.

Bennett, C. H. (1985), 'Dissipation, information, computational complexity and the definition of organization', in D. Pines (ed.), *Emerging Syntheses in Science*, London: Addison-Wesley, pp. 215–33.

Boccara, N. (2004), *Modeling Complex Systems*, New York: Springer-Verlag.

Bonabeau, E., Dorigo, M. and Theraulaz, G. (1999), *Swarm Intelligence: From Natural to Artificial Systems*, New York: Oxford University Press.

Brentano, F. (1995), *Psychology from an Empirical Standpoint*, New York: Routledge.

Cariani, P. (1991), 'Emergence and artificial life', in C. Langton, C. Taylor, J. D. Farmer and S. Rasmussen (eds), *Artificial Life II*, Redwood City, Calif. Addison-Wesley, pp. 775–97.

Casti, J. L. (1992), *Reality Rules: I Picturing the World in Mathematics – the Fundamentals*, New York: John Wiley.

Chaitin, G. J. (1966), 'On the length of programs for computing finite binary sequences', *Journal of the Association of Computing Machinery*, 13: 547–69.

Chaitin, G. J. (1997), *The Limits of Mathematics: A Course on Information Theory and Limits of Formal Reasoning*, Singapore: Springer-Verlag.

Chalmers, D. J. (2002), Foundations', in D. J. Chalmers (ed.), *Philosophy of Mind: Classical and Contemporary Readings*, New York: Oxford University Press, pp. 1–9.

Davidson, D. (1970), 'Mental events', in L. Foster and J. W. Swanson (eds), *Experience and Theory*, Amherst, Mass.: University of Massachusetts Press, pp. 79–101.

Dennett, D. C. (1987), *The Intentional Stance*, Cambridge, Mass.: The MIT Press.

Eberhart, R., Shi, Y. and Kennedy, J. (2001), *Swarm Intelligence*, San Francisco, Calif.: Morgan Kaufmann.

Ehresmann, C. (1968), 'Esquisses et types des structures algébriques', *Bulletin of Polytechnic Institute Iasi*, 14 (1–2): 1–14.

Eigen, M. and Schuster, P. (1979), *The Hypercycle: A Principle of Natural Self-Organization*, Heidelberg: Springer.

Fodor, J. A. (1975), *The Language of Thought*, Cambridge, Mass.: Harvard University Press.

Fodor, J. A. (1987), *Psychosemantics: The Problem of Meaning in the Philosophy of Mind*, Cambridge, Mass.: The MIT Press.

Fodor, J. A. (ed.) (1995), *Fodor, Jerry A.*, Blackwell Reference Online. Blackwell Publishing.

Gazzaniga, M. S. (1989), 'Organization of the human brain', *Science*, 245 (4921): 947–52.

Glasersfeld, E. von (1998), 'Anticipation in the constructivist theory of cognition', in D. M. Dubois (ed.), *Computing Anticipatory Systems*, Woodbury, NY: American Institute of Physics, pp. 38–47.

Haken, H. (1983), *Advanced Synergetics: Instability Hierarchies of Self-Organizing Systems and Devices*, Berlin: Springer-Verlag.

Harman, G. (1989), 'Some philosophical issues in cognitive science: qualia, intentionality, and the mind–body problem', in M. I. Posner (ed.), *Foundations of Cognitive Science*, Cambridge, Mass.: The MIT Press, pp. 831–48.

Horgan, T. (1993), 'From supervenience to superdupervenience: meeting the demands of a material world', *Mind*, 102 (408): 555–86.

Kauffman, S. (1969), 'Metabolic stability and epigenesis in randomly connected nets', *Journal of Theoretical Biology*, 22: 437–67.

Kelso, J. A. S. (1995), *Dynamic Patterns: The Self-Organization of Brain and Behavior*, Cambridge, Mass.: The MIT Press.

Kim, J. (1992), 'Multiple realization and the metaphysics of reduction', *Philosophy and Phenomenological Research*, 52 (1): 1–26.

Kolmogorov, A. (1965), 'Three approaches to the quantitive definition of information', *Problems of Information Transmission*, 1 (1): 1–17.

Langton, C. (1990), 'Computation at the edge of chaos: phase transitions and emergent computation', *Physica D*, 42 (1–3): 12–37.

Lawvere, F. W. (1963), 'Functorial semantics of algebraic theories', New York: Columbia University.

Le Corbusier (1985), *Towards a New Architecture*, New York: Dover.

Letelier, J.-C., J. Soto-Andrade, F. Guinez Abarzua, A. Cornish-Bowden and M. Luz Cardenas (2006), 'Organizational invariance and metabolic closure: Analysis in terms of (M,R) systems' *Journal of Theoretical Biology*, 238 (4): 949–961.

Mclaughlin, B. P. (1992), 'The rise and fall of British emergentism', in A. Beckermann, H. Flohr and J. Kim (eds), *Emergence or Reduction? Essays on the Prospects of Nonreductive Physicalism*, Berlin: Walter de Gruyter, pp. 49–93.

Mclaughlin, B. P. (1997), 'Emergence and supervenience', *Intellectica*, 25 (1): 25–43.

Millikan, R. G. (2004), *Varieties of Meaning: The 2002 Jean Nicod Lectures*, Cambridge, Mass.: The MIT Press.

Mitchell, W. J. (1990), *The Logic of Architecture: Design, Computation and Cognition*, Cambridge, Mass.: The MIT Press.

Nicolis, G. and Prigogine, I. (1967), 'On symmetry breaking instabilities in dissipative systems', *The Journal of Chemical Physics*, 46 (9): 3542–50.

Place, U. T. (1956), 'Is consciousness a brain process?', *British Journal of Psychology*, 47: 44–50.

Prigogine, I. and Stengers, I. (1984), *Order Out of Chaos: Man's Dialogue with Nature*, Toronto: Bantam Books.

Putnam, H. (1973), 'Psychological predicates', in W. H. Capitan and D. D. Merrill (eds), *Art, Mind and Religion*, Pittsburgh, Pa: University of Pittsburgh Press, pp. 37–48.

Rosen, R. (1991), *Life Itself: A Comprehensive Inquiry into the Nature, Origin, and Fabrication of Life*, New York: Columbia University Press.

Ryle, G. (1949), *The Concept of Mind*, London: Hutchinson.

Schuster, H. G. (2001), *Complex Adaptive Systems: An Introduction*, Saarbrücken: Scator Verlag.

Searle, J. R. (1983), *Intentionality: An Essay in the Philosophy of Mind*, Cambridge: Cambridge University Press.

Smart, J. J. C. (1959), 'Sensations and brain processes', *Philosophical Review*, 68: 141–56.

Smithers, T. (2002), 'Synthesis in design', in J. S. Gero (ed.), *Artificial Intelligence in Design*, Cambridge: Kluwer, pp. 3–24.

Sosa, R. and Gero, J. S. (2005), 'A computational study of creativity in design: the role of society', *Artificial Intelligence for Engineering Design, Analysis and Manufacturing*, 19 (4): 229–44.

Von Foerster, H. (1984), *Observing Systems*, Salinas, Calif.: Intersystems.

Wolfram, S. (1994), *Cellular Automata and Complexity*, Reading, Mass.: Addison-Wesley.

Zamenopoulos, T. and Alexiou, K. (2007), 'Towards an anticipatory view of design', *Design Studies*, 28 (4): 411–436.

This page intentionally left blank

7 The art of complex systems science

Karen Cham

1 Introduction

> at the nascent moment of creativity boundaries between disciplines dissolve
>
> (Miller 2000)

The scientific study of complex systems encompasses more than one theoretical framework, as it is has had to be highly interdisciplinary in seeking answers to fundamental questions about adaptable, changeable systems. Representing this knowledge in a usable way is one of the key ways of understanding complex systems behaviours and developing a workable complex systems science.

It is by means of representation that science has always apprehended, represented, understood and anticipated parts and processes of the world around us; indeed, when scientific representations have failed, as was the case for atomic physics in the 1920s, science has struggled to proceed. The standard methodology in scientific research has always been mathematics. Do the complex systems scientists of the future need to look for different representational methods?

It is in the arts that one finds expert knowledge and practices for innovation in representation; documentation, visualisation, simulation and embodiment are all artistic methods that can represent complex systems. Indeed, there are some precedents that will be expounded here that have thus far taken divergent perspectives – from Casti and Karlqvist's (2003) algorithmic complexity of art images to Galanter's (2003) notion of complexity theory as a context for art theory. However, it is also important to point out that modelling knowledge of complex states via illustration, diagrams and maps may be helpful, but to the art world those representations are not in themselves 'art'; they are the 'result of a mental effort' and are well accounted for as knowledge-elicitation methods in many disciplines, not least in systems science. So, if we are to argue that art processes and practices are to be useful to complexity science, we shall also have to provide a working definition of what makes a representation 'art' and, of course, what is 'not art'.

Conventionally, it seems that aesthetic knowledge and the use of metaphor have been under-valued, despite the key artistic process of drawing being well established as a means of problem-solving in the most 'scientific' of the arts such as architecture and engineering. It seems that cognitive processes that involve intuitive use of aesthetic languages are generally under-valued – somehow less useful or less *real* than logic and rationale. Is this because they cannot (yet?) be measured by mathematics? Intuitive thinking and aesthetic metaphor have long been established as a cogent part of rational processes; indeed, Einstein's concept of intuition included modes of mental imagery, and there is convincing evidence of the role of intuition in many scientific discoveries. This paper will seek to establish that aesthetic exploration of scientific subjects reaps tangible rewards, by citing key examples from contemporary medical research.

Finally, it is argued here that it is unsurprisingly within art theory that paradigms on aesthetic processes, practices and methodologies can be found to inform the development of pluralist ontologies and complex adaptive methodologies necessary to develop the art of complex systems science. This paper aims to set out the proposal that a poststructuralist approach can deliver ontological bases that will provide complex systems science not simply with diagrams and maps, nor even with the workable metaphors of visualisation, simulation and embodiment, but with visualisability – that most elusive and 'scientific' of representations that shares a generative semantic relation with that which it represents.

2 Complex systems and science

> Complex systems are an invention of the universe. It is not at all clear that science has an *a priori* primacy claim to the study of complex systems
>
> (Galanter 2003)

Academic dialogues have begun to explore the phenomenon of 'complexity' primarily as a new type of 'systems theory', an applied transdisciplinary methodology that finds its roots in Bertalanffy's 1969 'General Systems Theory'. Systems theory allows us to apprehend mechanical systems such as a car engine, biological systems such as the heart, or social systems such as a school from a common theoretical perspective. More broadly, 'systems thinking' is best-recognised as a dialectical method that breaks with logical and causal analyses to emphasise relationships within a whole. It can be traced from Socrates through Hegel to pragmatics as a means of identifying systemic principles common to different 'systems' from different perspectives.

In their concern with systems and relationships, both systems theory and complexity theory stand in stark contrast to conventional science which, based upon Descartes' reductionism, aims to analyse systems by reducing phenomena to their component parts (Wilson 1998). Reductionism is most easily understood in the context of the 'natural sciences' where the nature of complicated phenomena is understood by reducing them to a more fundamental form, for example, matter>molecule >atom>nucleus.

The scientific study of complex systems already encompasses more than one theoretical framework, as it has had to be highly interdisciplinary in seeking the answers to fundamental questions about living, adaptable, changeable systems. Thus there is no single unified Theory of Complexity; but several different theories have arisen from the natural sciences, mathematics and computing, and have been developed through artificial intelligence and robotics research, with other important contributions coming from thermodynamics, biology, physics, sociology, economics and law.

For our purpose here, 'complex systems' will be the general term used to describe those systems 'that are diverse and made up of multiple interdependent elements', that are often 'adaptive', in that they have the capacity to change and learn from events, and that can be understood as emerging from the interaction of autonomous agents – especially people (Johnson 2007).

Some generic principles of complex systems (Mittleton-Kelly 2003) that are of concern here are:

- self-organisation
- emergence
- interdependence
- feedback
- space of possibilities
- co-evolving
- creation of new order

In common terminology, describing something as 'complex' is often a point of resignation, a term that infers that something cannot be sufficiently assessed or controlled; supply-chain logistics, rail networks and organisational infrastructures are all commonly described in this way. In socio-cultural dialogues, the term 'complex' is used to describe humanistic systems that are 'intricate, involved, complicated, dynamic, multi-dimensional, interconnected systems (such as) transnational citizenship, communities, identities, multiple belongings, overlapping geographies and competing histories' (Cahir and James 2006).

Representing this knowledge in a usable way is one of the key methods of understanding complex systems behaviours. The notion that representations can facilitate understanding has long been debated. Certainly effective representation functions as a cognitive aid to understanding, and representing knowledge in a usable way is one of the key ways of problem-solving. Herbert Simon goes so far as to state that 'Solving a problem simply means representing it so as to make the solution transparent' (Simon 1996).

By means of effective representation it may be possible to develop scientific understanding of complex systems, or allow scientific understanding to be amended in the light of the behaviour of complex systems. The scientific study of complexity could underpin design and management strategies for industry, research and the marketplace. So it is not so much that the study of complexity 'should be a science', but rather what that scientific study can provide.

It is by means of accurate representation that science has always apprehended, represented, understood and sometimes successfully anticipated the behaviour of the parts and processes of the world around us. However, our knowledge of complex systems already seems to be too subtle for conventional approaches.

What happens when new phenomena need to be represented and they do not work according to pre-established concepts? As Miller (2000) sets out in his brilliant book *Insights of Genius: Imagery and Creativity in Science and Art*, in the early days of atomic physics, representational issues dogged research at every turn. For example, how could one represent something that is simultaneously wave and particle? At the atomic level, everything was different.

According to Miller, it was the collapse of Niels Bohr's hugely successful atomic theory that prompted the founder of quantum mechanics, Werner Heisenberg, to state that 'there can be no directly intuitive geometrical interpretation because the motion of electrons cannot be described in terms of the familiar concepts of space and time' (Miller 2000).

Between 1913 and 1923, Bohr's atomic theory rested on the intuitive and aesthetically logical notion that the hydrogen atom operated like a small planetary system, and quite some progress was made using this model. Unfortunately, when applied to the helium atom, it provided no useful insights at all, and Bohr's theory began to fall apart.

In 1949, after over twenty-five years of atomic physicists working without any workable visual imagery at all, Richard Feynman used the mathematical formalism of quantum mechanics to generate the proper visual imagery. This was not high-risk 'visualisation' but 'visualisability'. While the former is understood as having an abstract relation to that which it represents, the latter is more directly related to the properties of the subject: for example, the movement of iron filings in relation to a magnetic force. It was Feynman's visualisability that helped the representation of quantum states transcend the paradigms of space time.

For quantum theory, 'the concept of objective reality evaporated into the mathematics that represents no longer the behaviour of elementary particles but rather our knowledge of this behaviour' (Heisenberg 1958). Even Wittgenstein recognised that mathematics was simply a matter of inventing and applying concepts according to agreed practices. According to the Spanish scientist Jorge Wagensberg, 'if the nature of science were to be considered under the premises of reality, plausibility, and dialectics, then whoever attempted to identify these three principles by strictly observing the complexity of the objects would reach the conclusion that the object resisted the method. The only manner of proceeding would be to 'soften up' the method, with the result that science is transformed into ideology (Giannetti 2007). Feyerabend, too, asserts that science is most definitely in the proximity of art, as he negates the possibility of absolute rationality and logic in regard to anything which is created by the human mind. Thus the sciences are not an institution of objective truth, but are arts along the lines of a progressive understanding (Giannetti 2007).

In the arts, this is known as the 'postmodern' condition – a much maligned and misrepresented term that serves to describe a loss of faith in the self-legitimising grand narratives of logic and reason and the recognition that all knowing is narrative knowing, and all schemata express not the nature of 'reality' but the nature of mind. It is poststructuralist theory that provides the critical tools for the postmodern condition.

Can poststructuralist art theory provide ontologies that can generate complex systems; that is, make complex systems visualisable?

3 Art and science

> good mathematicians see analogies between theorems. The very best ones see analogies between analogies.
>
> (Miller 2000)

Prior to the Age of Enlightenment, the work of the scientist and the artist were often indistinguishable; one only has to think of the work of Leonardo Da Vinci or Albrecht Dürer to perceive the unity of art and science at that time.

Indeed, scientific principles have always underpinned art production: the rules of perspective, proportion and the golden section for example. In Mick Wilson's paper 'How should we speak about art and technology?' (2001), he describes the radical separation of art and science as a recent phenomenon, 'enmeshed within the complex historical process of modernisation'. He cites Kristella's demonstration that the (fine) arts *per se* were constituted as a separate arena of human endeavour only as late as the seventeenth or eighteenth centuries. It is actually even later, when the term 'art' becomes 'Art', associated with creativity, expression, the affective and subjective, that the practice becomes understood as diametrically opposed to the sciences (Wilson 2001).

Cognitive processes that involve intuitive use of aesthetics generally became under-valued – seen as somehow less useful or less *real* than logic and rationale, despite intuitive thinking and visualisation being a cogent part of rational processes. There is, for example, convincing evidence of the role of intuitive visualisation in many scientific discoveries (Jorgensen 2006), while some fundamental connection between intuitive knowing and aesthetics is enshrined in Einstein's concept of intuition (Miller 2000). It is thus in the arts that one now finds expert knowledge and practices for innovation in representation. Documentation, visualisation, simulation and embodiment are all artistic methods that can represent indefinables, randomness and chaos. As Anders Karlqvist puts it, 'art tries to capture . . . what cannot be represented in a formalized scientific mode' (Casti and Karlqvist 2003).

There are also excellent contemporary precedents where the aesthetic work of artists is contributing to the research work of scientists in tangible and useful ways. For example, Dr Ian Thompson, Research Fellow in the Department of Oral and Maxillofacial Surgery, at Guy's Hospital, King's College London, works with artist Paddy Hartley on the 'Face Corsets and Bioactive Glass Facial Implants Project' funded by a Wellcome Trust People Award. Paddy has devised face 'corsets' as works of art that have gone on to be adapted for medical application. Here, the exploratory work of the artist has informed the development of new medical practices.

The work of Jane Prophet, too, has been seminal in establishing the role of aesthetic knowledge in informing science. Her rapid prototyping of the human heart was of great interest to heart surgeons who had never experienced a human heart in three dimensions; when one opens up the chest the change in pressure causes the heart to collapse. Through her sculptures, Jane had created a unique learning opportunity for a medical scientist.

In CELL (2002–) Jane collaborated with a mathematician, a scientist, a stem cell researcher, an Artificial life programmer, and a curator/producer to explore ideas about how adult stem cells behave and organise themselves. Research outputs included a formal mathematical model, simulations, visualisations and a series of art pieces that have resulted 'from looking at the overall nature of our combined multi-disciplinary attempt to investigate new theories of biological organisation' (D'Inverno and Prophet 2004).

In her book *Artists-in-Labs: Processes of Inquiry* (2006), Jill Scott argues that it is essential for the arts and the sciences to work together to develop holistic approaches to innovation and representation. She defines the key aims of her Artists-in-Labs project as follows:

- to increase the levels of know-how transfer between art and science and further the potentials for collaboration between artists and scientists
- to give artists the opportunities to comment and reflect upon the implications and results of scientific research in relation to society
- to encourage the innovation potential of Swiss artists in the fields of new technological developments and applications as well as the potential to gain technical and scientific expertise
- to act as an agent between art and science in the public realm and further the understanding of each other's methodologies of production

In the arts and the humanities, methodological approaches to the analysis of complex systems can be traced back to the work of Ferdinand de Saussure in the early part of the twentieth century. Saussure's unprecedented exploration of the notion of language addressed the general structural properties of language *itself* rather than the attributes of specific languages. Derivative of Saussurian linguistics, structuralism and its progenesis, poststructuralism, underpins a huge amount of methodologies and practices which extrapolate well across divergent cultural forms and practices from literature, anthropology, art, design and media, and account for complex behaviours (Cilliers 1998).

Saussure's work provided an ontology of language as a primary complex system, over fifty years before such a perspective was common in other disciplines;

his notion of *langue*, the system of rules, and *parole*, the performative aspects of lan guage in use, is a seminal model for understanding the nature of complex systems. Indeed, linguistics has long argued that it is the *scientific* study of language and that semiotics is the *science* of sign usage within linguistic structures. If interaction is key to complex behaviour, surely it is possible to apply this body of theory about interaction to the study of complexity?

4 Art and complexity

> The science of complex systems is still very young – and there are many different views on what constitutes this science. Art may generate new ideas and help to solve problems, may give means of communicating complexity, and may provide new methods of scientific inquiry.
>
> (Johnson 2007)

There is an established, if divergent and sporadic, history of exploring art in relation to complex systems (Casti and Karlqvist 2003). It is also not unusual to find artworks themselves described as complex systems. In 2002, the Samuel Dorsky Museum of Art, New York, held an international exhibition entitled *Complexity; Art & Complex Systems* that was concerned with 'art as a distinct discipline offer(ing) its own unique approache(s) and epistemic standards in the consideration of complexity' (Galanter and Levy 2002: 5).

The exhibition showed the work of twenty-four artists, including: Mauro Annunziato, who investigates scientific complexity and artificial life; Paul Hertzcreates, whose prints and installations implement a 'competitive cellular automata mechanism' which results in a self-organising pattern; and Manuel Baez, who explores form, structure and process to generate emergent properties.

The organisers summarise four ways in which artists engage with the realm of complexity:

- presentations of natural complex phenomena that transcend conventional scientific visualisation
- descriptive systems which describe complex systems in an innovative and often idiosyncratic way
- commentary on complexity science itself
- technical applications of genetic algorithms, neural networks and a-life

In 1998, the Swedish research agency FRN sponsored a one-week workshop to bring together complexity scientists interested in art and artists interested in complexity to explore how the two subjects overlapped, complimented and contrasted with each other. Subjects covered at this seminal event were as divergent as the Internet, birdsong, fractals and drawing.

In his paper, John Casti asked whether 'good art is complex art' and explored the notion of an 'algorithmic complexity'. Algorithmic complexity is one of the most well established approaches to characterising the complexity of the visual image and can be traced back to Garrett Birkhoff's 'Aesthetic Measure' (Casti and Karlqvist 2003) of the 1930s. Birkhoff's work involved developing equations to account for what he termed the 'order' versus 'complexity measure'. His interest was in finding an equation for aesthetic harmony, and he went on to explore works of art from a geometrical perspective, assessing relationships within the composition and the 'chiaroscuro', the correct term for the balance of contrast between light and dark in an artwork.

Casti extrapolates Birkhoff's notion of complexity in the image to define his notion of algorithmic complexity as measured by the length of the shortest computer program that could reproduce it. Figure 7.1, for instance, demonstrates low, medium and high algorithmic complexity of an image. Obviously certain variables, such as the processing power of the machine and the language used, are anchored for consistency. There is a huge amount of research in this area with little, if no, penetrable insights into artworks. Casti concludes that a syntactic approach to measuring the complexity of a work of art is simply not complex enough. Measuring the syntax of the image generates a result that is inane. This not only demonstrates that Birkhoff's mathematical approach to measuring 'aesthetics' is actually only applicable to the complexities of mathematical proportions, geometries and ratios in composition; tt also demonstrates that the complexities of the image lie in its *semantics*. Failing to account for semantics is failing to account for complexity.

The advent of computer-generated art, too, has led to predictable results for algorithmic complexity. High-complexity images result in patterns which are semantically very simple.

Processing power has facilitated a generation of highly algorithmically complex images like Figure 7.2. However, while it is attractive, it demonstrates little more than mathematical processes and their semantic simplicity. Concurrently, computational modelling of the complexity of natural phenomena has also reached a degree of velocity. 'Many beautiful and intricate mathematical patterns – regular geometric shapes, fractal patterns, the laws of quantum mechanics, elementary particles,

Figure 7.1 Low, medium and high algorithmic complexity of images. Source: Mark Dow

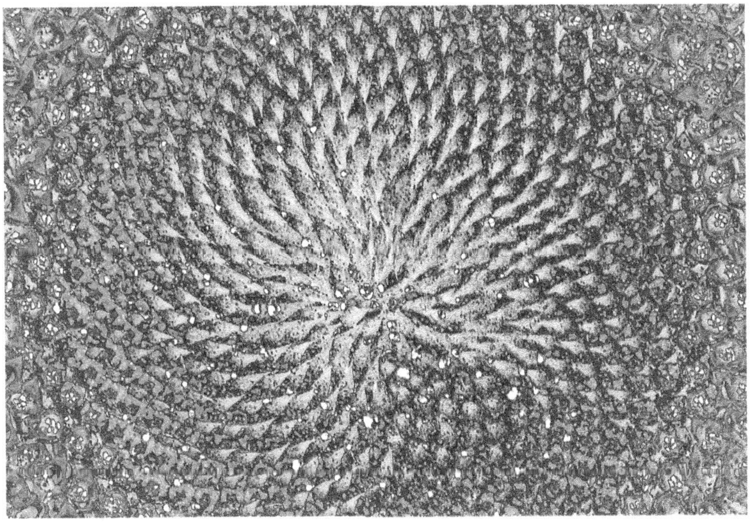

Figure 7.2 'Galactic', algorithmic art, vector art, flower photography, by Stephen Linhart, 2006. Source: http://www.stephen.com/. Image courtesy of the artist

the laws of chemistry – can be produced by short computer programs.' This serves finally to anchor Birkhoff's notion of aesthetic measure, and that approach to algorithmic complexity, in the realm of geometry and the physical sciences rather than in that of semantics and culture.

So we can demonstrate an aesthetic model of mathematical shapes and algorithmical models of nature, yet we are no nearer to assessing the complexity of art. As part of the FRN project exploring art and complex systems, Bobo Hjorts' paper focused upon drawing as an evaluative process – a type of formulated knowledge that uses visual/intuitive thinking. He proposes that 'a scientist has little benefit of meeting with artists, compared to what he can reach by learning and practicing the *method* of artists . . . how scientists can learn from, and make better use of the artists' way of thinking' (Casti and Karlqvist 2003).

If we take as an example Zen calligraphy, we can perhaps understand a little more about cultural artefacts and their relationship to complexity theory. The Zen character could perhaps be understood as having a low aesthetic measure or algorithmic complexity, as the Zen Master strives to produce the image with as few strokes as possible; Zen is the art of simplicity.

The Zen character, however, is not 'simple' in the way in which a Western calligraphic character might be. For all its 'algorithmic complexity', Figure 7.3 only represents the phonetic sound 'sss'. While the typeface and scroll pattern have specific semantic connotations for the graphic designer, it is still a patterned letter.

The Zen character, however, is a type of symbolic hieroglyph in that it manifests complex meanings in minimal form (Figure 7.4). 'During the writing, body and mind leave their traces on the paper. An ink trace or Zen calligraphy is created in a single moment, in one breath. Ink traces not only show the form of the kanji and express the meaning of it. They are also a living document of an intensively experienced moment' (Kuwahara 2006).

Figure 7.3 Western calligraphy – the letter s. Source: Denise Sharp (studiodsharp.com)

Figure 7.4 Zen character. Kokoro, Heart, by Kuwahara, 2006. Source: Kokugyo Kuwahara

How might the notion of complexity apply to a Zen character? The Zen mark is aesthetically 'simple' in that it is made of only four marks; but, from the description of the process, it is in fact an emergent property of the complex interaction between the master's 'body and mind', his tools, the concept of 'heart' and the materials. The next time it is made, it may be similar; but, like a snowflake, every Zen kanji is unique.

While one might argue that every freehand illustration is to some extent an emergent property of a complex interaction, 'a living document . . . of a moment', what is important here is that the Zen hieroglyph is the very opposite of the algorithmic art of Figure 7.2; it is an aesthetically simple, emergent property of a complex semantic process. Figure 7.2 is a semantically simple, generative result of an aesthetically complex process. In the Zen kanji, perhaps the algorithmic complexity of the aesthetics could be accounted for successfully, but the complexity of the *meaning*, anchored in the process, would remain extraneous to the equation.

The intricacies of the relationships between aesthetics and semantics are not for exploration here; suffice it to say that perhaps the most important insight into the complexities of artistic processes may be first to recognise that the process of 'making a mark' and the 'mark' itself are always inextricably linked both to each other and to a *context*. It is only within these relationships that the semantics of visual representation can be observed; and, as we have discovered, it is within the semantics that we can observe complex behaviours.

If the study of complex systems is to develop into a true science, then these 'semantic behaviours', as we can perhaps now term them, must meet the criteria of 'scientific reproducibility', one of the main principles of the scientific method. This refers to the ability of a test or experiment to be accurately reproduced, or replicated, by separate, independent working; and thus scientific reproducibility requires a recognisable, definable and transferable *method*. It is on this basis that the scientific approach established the validity of a representational form.

It is in semiotics, the study of the science of signs (Cobley and Jansz 1997), that all signs are recognised as demonstrating a degree of complexity. The fluid relationships between the signifier (the mark), the signified (the attendant concept), the reader or 'interpretant' (Cobley and Jansz 1997) and the context within which it is read are understood as being in a state of constant relational flux. Poststructuralist semiotics recognises that the given meaning of any sign is therefore never fixed, but co-evolving; that the meaning of any sign at any given moment is an emergent property of the co-evolving ecosystem of nested complex systems of the sign in use.

5 Art in the science of complex systems

> in the coming decades, the economies of the western world will increasingly deal with communication, experience, entertainment, messages, visualisations etc. The real experts in all this are the artists.
>
> (Casti and Karlqvist 2003)

In 2007, a three-day research symposium and workshop 'Art in the Science of Complex Systems' was convened by Jeffrey Johnson, Professor of Complexity Science and Design at The Open University and Chair of the Complex Systems Society, in collaboration with the University of Brighton, to explore the specific research question of 'what contribution the arts might make to the *science* of complex systems'. Contributors were specifically asked to answer this question from an academic research perspective as part of the AHRC/EPSRC-funded 'Embracing Complexity in Design' research cluster.

An international, interdisciplinary group of artists, scientists, mathematicians, musicians, poets and performers were invited to explore the question from an academic research perspective. Guests included John L. Casti, Professor of Operations Research and System Theory at the Technical University of Vienna and a member of the External Faculty of the Santa Fe Institute, New Mexico; Paul Brown, Visiting Professor at the Centre for Computational Neuroscience and Robotics and Department of Informatics, University of Sussex and Chair of the Computer Arts Society; and Professor

Lizbeth Goodman, Director of the SMARTlab Digital Media Institute and Magic Gamelab at University of East London. All contributors were given an open opportunity to describe, present, install, perform, screen, sing or dance their response to the question.

The instigation for this event was Eve Mittleton-Kelly's work with ECiD artist Julian Burton in her pursuit of the development of a theory of complex social systems. Julian Burton makes work that visualises how companies operate in specific relation to their approach to change and innovation (Figure 7.5). He is a strategic artist and facilitator who makes 'pictures of problems to help people talk about them' (Burton, 2007). Clients include public- and private-sector organisations such as Barclays, Shell, Prudential, KPMG and the NHS. He is quoted as saying:

> Pictures are a powerful way to engage and focus a group's attention on crucial issues and challenges, and enable them to grasp complex situations quickly. I try and create visual catalysts that capture the major themes of a workshop, meeting or strategy and re-present them in an engaging way to provoke lively conversations.

Burton's work is a simple and direct method of using art as a knowledge elicitation tool that falls into the first and second categories of Galanter and Levy's summary of how artists engage with complexity, but are these images art? Modelling knowledge of complex states via illustration, diagrams and maps may be helpful; but, to the art world, those representations are not, in themselves 'art' – they are the 'result of a mental effort' and are well accounted for as knowledge-elicitation methods in many disciplines, not least in systems science. So, to argue that art processes and practices

© 2006 Julian Burton

Figure 7.5 'A picture speaks a thousand words', by Julian Burton, 2006. Source: the artist

are to be useful to complexity science, we must provide a working definition of what makes a representation 'art' and, of course, what is 'not art'.

An artwork is different from illustrations, maps and diagrams. An artwork is not the result of a logical analysis of a subject; it is a different type of representation; it is an intuitive aesthetic exploration of a subject – as Bobo Hjorts suggested at the FRN workshop, art is 'formulated knowledge using visual/intuitive thinking' (Casti and Karlqvist 2003). It is easier to understand this difference in terms of music; the semantic relation between Gustav Holst's Planets Suite and our idea of the solar system is different from the semantic relation between star maps, models of terrain, Hubble photographs and our scientific knowledge of the heavens. Both forms of representation tell us something about the subject, yet only the music would constitute art.

Additionally, a work of art can be said to 'simultaneously define and meet its own criteria'[1] in the sense that it holds within it its own transferable context, a type of self-containment that allows it to be experienced again and again in different times and places without losing resonance – a type of 'artistic reproducibility' perhaps.

It is here proposed, however, that *good* art is different in a third way from aesthetic exploration *per se*. Good art is different, first, in its semantic relation to its subject and, second, in its capacity to transform our pre-existing perceptions, understanding or experience of a subject *in an unpredictable, idiosyncratic way*. In this way, an artist is always concerned with facilitating emergence; an artwork is designed specifically to facilitate emergent interpretations (of a subject). Pablo Picasso described art as 'a lie which makes us realise truth' (Ashton 1988). The artwork is a conceptual 'participatory space' that enables engagement and interpretation, and is recognised in advertising design as essential to success (Plummer 2007). While advertisements can be said to engineer the emergence of semantic behaviours, they are not art because they are the result of logical analysis.

A working definition of a work of art for our purposes here is therefore tri-part:

- art is the formulated knowledge of aesthetic/intuitive thinking (after Hjorts)
- art simultaneously defines and meets its own criteria (after Scriven)
- art enables emergence

The ECiD event was similar to the FRN project of 1998 in embracing divergent artform practices. For example, Mick Wallis ran an introductory participatory workshop on 'Boalian bodywork' based upon the work of August Boal and his 'Boal Method of Theatre and Therapy' (Boal 1995) which explored the use of movement, sound and light to explore representation of complex phenomena. His work with Arts Work With People has a developed methodology to 'assist people (especially with severe access needs) to find artistic metaphors for their own complex relations with the world'. Wallis's workshop aimed to explore the application of this methodology to the articulation of scientific complexity from the point of view of the scientist.

At the other end of the arts spectrum, Carol Bellard Thomson's poetry readings explicated the value of aesthetic knowledge and metaphor, while her presentation

'Colourless Green Ideas Sleep Furiously' suggested that linguistics could provide complex systems science with workable ontologies. As an expert in linguistic stylistics, Carol asserted that linguistics can account for the multiple operational features and processes of the language system, and their interaction to produce meaning(s).

> *Objectivity – C. A. Bellard Thomson*
>
> The words wore out.
> All the delicate and fragile ones
> that could build filigreed elegance,
> all the thundering torrents
> to sweep away innocence,
> and all the gentle rocking words
> of motherhood and safety.
> All that was left was plain speaking,
> a Pied Piper in a threadbare coat,
> with an aching need for grace.

The author designed a participatory sculpture in collaboration with Jeffrey Johnson to serve as documentation of the time-based dialogic process. During the three-day workshop/exhibition the invited contributors and guests were able to contribute to the construction of the sculpture based on the aesthetics of scientific models of molecular structures. The title of the work, 'Matmos', was taken from the seething lake of evil slime beneath the city Sogo in the film *Barbarella* (Roger Vadim 1968). Here it is used as pun on the Cartesian division between mind and matter, the judeo-christian evil being, of course, anchored in the latter. Juxtaposed with the aesthetics of molecular structures, and the embodied process of making, the whole becomes a pun on Enlightenment thinking that is discombobulated by complexity (Cham and Johnson 2007).

As a participatory collaborative process, 'Matmos' (Figure 7.6) was designed by the author and Jeffrey Johnson, but made by the numerous attendees of the event. In this way it is definable as 'Systems Art'. 'Open Systems: Rethinking Art *c.*1970', at Tate Modern, London (2005), described system art as a 'rejection of art's traditional focus on the object, to wide-ranging experiments with media that included dance, performance and . . . film & video' (De Salvo 2005: 3). Artists included Andy Warhol, Richard Long, Gilbert and George, Sol Lewitt, Eva Hesse and Bruce Nauman.

For example, Warhol's Marilyn Monroe screen prints are systems art because 'the artistic value of the work, which traditionally lies in the execution of the object, is here in the relation between the media, the technique, the subject and the context' (Cham 2007).

Francis Halsall defines systems art from a systems theoretical perspective as

> emerging in the 1960s and 1970s as a new paradigm in artistic practice . . . displaying an interest in the aesthetics of networks, the exploitation of new

Figure 7.6 'Matmos' – Karen Cham, Jeffrey Johnson, aided and assisted by Vernon S. Partello, MD Consultant, Molecular Model Co.

> technology and new media, unstable or de-materialised physicality, the prioritising of non-visual aspects, and an engagement (often politicised) with the institutional systems of support (such as the gallery, discourse, or the market) within which it occurs.
>
> (Halsall 2005)

In this sense, once *applied* back in the organisation where emergent change can occur because of his work, Julian Burton's 'Art of Change' project could be described as Systems Art; the drawings themselves are 'not art', as they do not stand alone, but are an aesthetic part of the artistic process/conceptual system that constitutes the work.

Systems art shares it roots with 'poststructuralism'; the critical discourse from the arts that can account for both 'open' (Eco 1989) system of structural relations and 'death of the author' (Barthes 1978); where the 'reader' is credited with an active part in emergent, multiple and evolving interpretations of a cultural artefact.

It is clear that the theoretical and methodological relations between complexity science and cultural systems can be expounded especially well through poststructuralism. The work of Barthes, Derrida and Foucault offers us the notion of all cultural artefacts, that is *all* products and processes of human interaction, as complex systems of signs, whose meanings are not fixed but rather sustained by networks of relationships. Can poststructuralism provide ontologies to underpin a mathematical account of emergent semantic behaviour?

6 Digitally interactive art and complexity

It is surely in the convergence of art and design with computational algorithms that we might find the most useful processes and practices for complexity science.

> It is in design for interactive media, where algorithms meet graphics, where the user can interact, adapt and amend, that self-organisation, emergence, interdependence, feedback, the space of possibilities, co-evolution and the creation of new order are embraced on a day to day basis by designers and users alike.
>
> (Cham and Johnson 2007)

Digitally interactive media is a recent development and is defined here as 'a machine system which reacts in the moment by virtue of automated reasoning based on data from its sensory apparatus' (Penny 1996).

> Interactivity is most commonly an attribute of server based multimedia on the internet and is a specific attribute of digital media, although interactive systems are not necessarily screen based.
>
> (Cham 2007)

Interactive systems, however, all demonstrate complex behaviours.

A digitally interactive environment such as the Internet demonstrates many key aspects of a complex system. Indeed, it has already been described as a 'complexity machine' (Qvortup 2006). Digital interaction of any sort is thus a graphic model of the complex process of communication. Here, complexity does not need deconstructing, representing or modelling, as the aesthetics (as in apprehended by the senses) of the graphical user interface conveniently come first.

> When designing digitally interactive artifacts we design parameters or co ordinates to define the space within which a conceptual process will take place. We can never begin to predict precisely what those processes might become through interaction, emergence and self organisation, but we can establish conceptual parameters that guide and delineate the space of possibilities.
>
> (Cham 2007)

If complex behaviours result from any interactive system, how can we provide an insight for complexity science from digitally interactive art?

Christiane Paul (2003) defines a number of thematic trends in digital art:

- Artificial life and intelligent agents
- Telepresence, telematics and telerobotics
- Body and identity
- Data visualisation and mapping
- Text and narrative environments
- Tactical media and hacktivism

Galanter and Levy (2002, 2003) concurrently point out how artists' engagement with complexity is via 'technical applications of genetic algorithms, neural networks and a-life'. Digital arts have explored this potential from the earliest points that readily affordable technologies allowed it, and it has always been commensurate with complexity.

In 1995, Roy Stringer demonstrated something he had authored using basic artificial intelligence concepts and he was very excited to show the author how self-organisation and emergent behaviours constituted 'evolution' in his little digital critters. It was very much along the lines of Jane Prophet's 'Technosphere' of the same year, an online digital environment which attracted over a 100,000 users who created over a million 'creatures'.

TechnoSphere constituted one of the first 'virtual worlds' populated by artificial life forms on the Web. The digital ecology of the 3D world, which was housed on a server, was designed for participation; it was digital systems art, which at the time transformed the user's sense of human and technological systems. Such examples are now commonplace and therefore may have lost some of their transformative capacity.

While Technosphere explores artifical life concepts in a digital environment, Troy Innocent's lifeSigns applies linguistic principles. This work evolves multiple digital media languages 'expressed as a virtual world – through form, structure, colour, sound, motion, surface and behaviour' (Innocent 2007).

The work explores the idea of 'emergent language through play – the idea that new meanings may be generated through interaction between human and digital agents' (Innocent 2007). Thus this artwork combines three areas of converging research: artificial life, computational semiotics, and digital games. In Toshio Iwai's work, the emphasis is upon the simplicity of the structure that facilitates complexity.

In 'Modeling complex systems for interactive art' (2000), Sommerer and Mignonneau work in reverse, applying the concepts of complex systems to the creation of a generative system, an 'interactive, computer generated and audience participatory' artwork. Here we have a simulation of complex principles as art. So art can be complex, and complexity can be art. Where do we go from here?

Digital artists are consciously exploring complexity, not because they have studied complexity science, but because it is a *quality of their material*; digital art is 'the formulated knowledge of aesthetic/intuitive thinking' using 'complex media' (Cham 2007). If poststructuralism had not come first, we would have to invent it to analyse this phenomenon as a 'complex performative mediated process' (Cham 2007).

It is by predicating the active role of the 'reader' and their participation in 'open' works that poststructuralism anticipates the complexity of digital media. For example, in looking at non-interactive artworks, such as paintings and sculptures, the 'reader' experiences an individualised *interpretation*; in interacting with a digital installation, such as Camille Utterback's 'Text Rain' (Figure 7.7) where the dynamic system responds to motion, the user experiences an individualised *interaction*.

The theoretical and methodological relations between complexity science and digital culture can clearly be expounded especially well through poststructuralism.

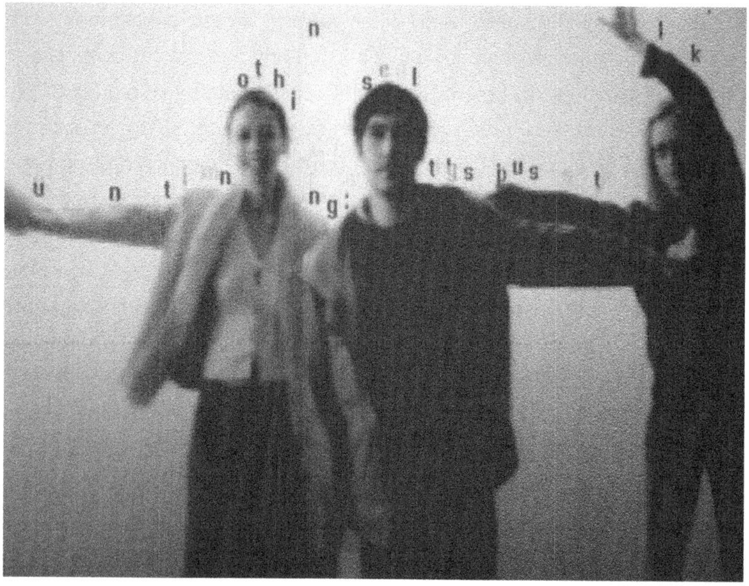

Figure 7.7 'Text rain', interactive installation by Camille Utterback and Romy Achituv, 1999. Source: the artists

> If language for Derrida is 'an unfixed system of traces and differences . . . regardless of the intent of the authored texts . . . with multiple equally legitimate meanings' (Galanter 2003) then I have heard no better description of the signifiers, signifieds, connotations and denotations of digital culture.
>
> (Cham 2007)

Implemented in a digital environment, poststructuralist theory is tangible complexity.

The potential integration of complexity and poststructuralist theory is explored theoretically in Paul Cilliers' 1998 book *Complexity and Postmodernism* where the author proposes an understanding of both complexity and computational theory through poststructuralism; a more in-depth philosophical treatise can be found in 'Foucault as complexity theorist' by Mark Olsen (2008). Arguably, this type of 'critical theory' provides ample functional methodologies for scientific modelling (Landow 1992), yet even in areas where a convergence of humanities methodology with scientific practice might seem to be most pertinent most examples are few and far between. In his paper 'Post structuralism, hypertext and the World Wide Web', Luke Tredennick states that 'despite the concentration of post-structuralism on text and texts, (even) the study of information has largely failed to exploit post-structuralist theory' (Tredennick 2006).

What might a digital model of the poststructuralist 'open text' tell us about designing for emergence in transformative representational systems? As a tentative step towards exploring the relationship between digital arts, poststructuralism and complexity, the author has, in collaboration with Dr Trevor Collins of The Open University's Knowledge Media Institute, devised a computational media semantics prototype based upon Roland Barthes' seminal analysis of the Panzani Pasta advertisement in his essay

'Rhetoric of the image' (Barthes 1978). 'Computing Italianicity' is a computational system based upon the ontological analysis of food advertising provided by Barthes' original work (Figure 7.8).

Like Barthes' treatise, the system is based upon the juxtapositioning of individual elements, with the initial system itself suggesting semantically logical options, derivative of certain user combinations. However, the system allows user tagging, so through use it is possible to begin to evidence the consensual visual language of media culture as it evolves.

What is important here is that it is a work of systems art, which, it could be argued, provides 'visualisability' of complex processes through the application of poststructuralist theory to the design of computational systems. This marrying of cultural theory and digital practice is core to the development of complex systems science; it is in digital discourse that one can most easily observe complexity in action. In digital culture, art and mathematics converge into a 'process driven, performative event (that) demonstrates emergence through autopoietic processes' (Cham and Johnson 2007).

Indeed, there is also an interdisciplinary concern with the potential of 'cultural science' that marries complexity and cultural studies.

> Two knowledge movements, cultural studies with roots in the humanities and complexity studies in the sciences have challenged the separation of the sciences, social sciences, and the humanities by upsetting the mutually exclusive epistemologies . . . in knowledge production.
>
> (Lee 2007)

Without recourse to poststructuralist analysis of culture, complexity science risks working blind.

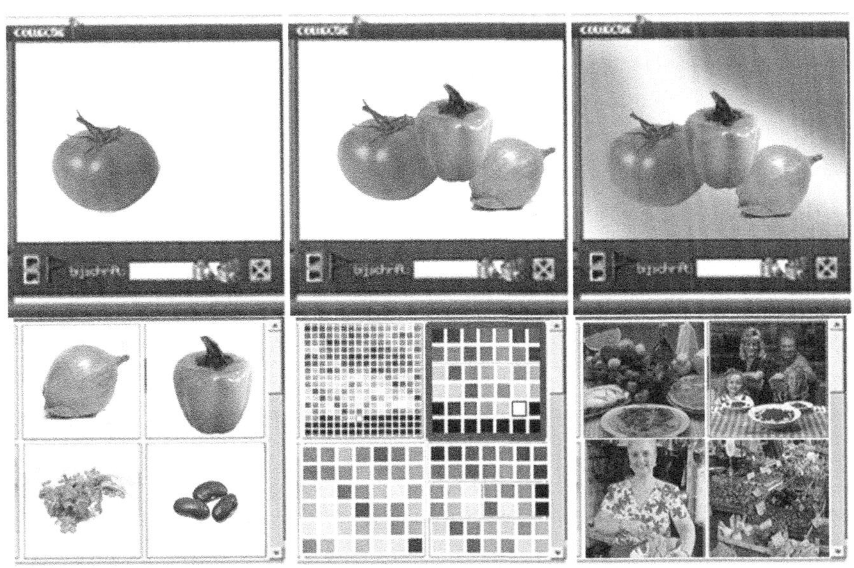

Figure 7.8 'Computing Italianicity' – Karen Cham and Trevor Collins, 2006

7 Conclusion

We have seen that representation is crucial to developing scientific insights, and that representations sometimes provide more than can be anticipated; indeed, it could be argued that they *participate* in the insights they afford us. Furthermore, it has been established not only that arts and sciences share common roots and aims, but also that there is a growing and legitimate practice of using arts practices to inform scientific developments.

We have also argued that in developing the science of complex systems it is crucial to recognise that complex systems display cultural characteristics – the point at which the polarised concepts of 'artificial' and 'natural' converge – so, to design for complexity, we must design for culture. It is thus no surprise, then, that the arts harbour people, processes and practices that can inform the science of complex systems.

We have demonstrated that art is a highly complex process demonstrating a complex engagement with the co-evolving nested ecosystems of culture – a dialogic participation that, at best, engineers emergent perceptions, interpretations and meanings. Furthermore, we have also suggested how, in applied art practices such as advertising design, simplified cultural artefacts such as *commodities* are anchored to specific complex messages with the aim of engineering *semantic behaviour*, and that the theoretical dialogue that accounts for all of these cultural artefacts, practices and contexts is poststructuralism.

It has also been demonstrated that Systems Art is systems thinking in art practices – an *explicated* dialogic approach to cultural systems that engages with the forms of popular culture. So, in conclusion, it is argued that, while using art as a knowledge-elicitation tool can take many forms, from creative visualisation to simulation and embodiment, it is systems art and poststructuralism that can provide complex systems science not simply with diagrams and maps, nor even with the workable metaphors of visualisation, simulation and embodiment, but with *visualisability* – that most elusive and 'scientific' of representations that shares a *generative* semantic relation with that which it represents, because that is what digital artists do.

Note

1 Robert Scriven, sculptor, in conversation with the author, 1988.

References

Ashton, D. (1988), *Picasso on Art*, New York: Da Capo.
Barthes, R. (1978), *Image–Music–Text*, New York: Hill & Wang.
Bertalanffy, L. von (1969), *General System Theory: Foundations, Development, Applications*, New York: Braziller.

Boal, A. (1995), *The Rainbow of Desire: The Boal Method of Theatre and Therapy*, London: Routledge.

Burton, J. (2007), Hedron People. 9 March. http://www.hedron.com/network/assoc.php4?associate_id=14

Cahir, J., and James, S. (2006), 'Complex: call for papers', *M/C Journal* 9 September. http://journal.mediaculture.org.au/journal/upcoming.php.

Casti, J. and Karlqvist, A. (eds) (2003), *Art and Complexity*, London: Elsevier.

Cham, K. and Johnson, J. H. (2007), 'Art in the science of complex systems'. http://artandcomplexity.wordpress.com/exhibition/

Cham, K. L. (2007), 'Reconstruction theory: designing the space of possibility in complex media'. Special Issue: Performance and Play: technologies of presence in performance, gaming and experience design, *International Journal of Performance Arts & Digital Media*, 2–3 (3), ed. Lizbeth Goodman, Deeverill, Esther MacCallum-Stewart and Alec Robertson Cham, K. L. and Johnson, J. H. (2007), 'Complexity theory: a science of cultural systems?' *M/C Journal*, Complex, 10 (3). Ed. J. Cahir and S. James, http://journal.media-culture.org

Cilliers, P. (1998), *Complexity and Postmodernism: Understanding Complex Systems*, London: Routledge.

Cobley, P. and Jansz, L. (1997), *Introducing Semiotics*, Royston: Icon.

De Salvo, Donna (2005), *Open Systems: Rethinking Art c.1970*, London: Tate Gallery.

d'Inverno. M. and Prophet, J. (2004), 'Modelling, simulation and visualization of adult stem cells', in P. Gonzalez, E. Merelli and A. Omicini (eds), *Fourth International Workshop on Network Tools and Applications*, NETTAB, pp. 105–16.

Eco, E. (1989), *The Open Work*, Cambridge, Mass.: Harvard University Press.

Galanter, P. (2003), *Against Reductionism: Science, Complexity, Art and Complexity Studies*, http://isce.edu/ISCE_Group_Site/web-content/ISCE_Events/Norwood_2002/Norwood_2002_Papers/Galanter.pdf

Galanter, P. and Levy, E. (2002), *Complexity: Art and Complex Systems*, New Paltz, NY: Samuel Dorksy Museum of Art, www.newpaltz.edu/museum/exhibitions/complexity.pdf

Galanter, P. and Levy, E. K. (2003), 'Complexity', *Leonardo*, 36 (4): 259–60.

Giannetti, C. (2007), *Art, Science and Technology*, http://www.medienkunstnetz.de/themes/aesthetics_of_the_digital/art_science_technology/

Halsall, F. (2005), 'Observing "Systems-Art" from a Systems-Theoretical Perspective', CHArt.

Heisenberg, W. (1958). *Physics and Philosophy: The Revolution in Modern Science*, London: Collins.

Innocent, T. (2007), 'Life Signs', 9 March, <http://www.iconica.org/main.htm>.

Johnson, J. (2007), 'Embracing complexity in design (ECiD)', http://www.complexityanddesign.net/

Jorgensen (2006), 'Towards a research agenda for visual informatics', *Proceedings of the American Society for Information Science and Technology*, 42 (1).

Kuwahara, K. (2006), online at www.doku-zen.de/en/bokuseki/bokuseki_einf.html

Landow, G. P. (1992), *Hypertext: The Convergence of Contemporary Critical Theory and Technology*, Baltimore, Md: Johns Hopkins University Press.

Lee, R. E. (2007), 'Cultural studies, complexity studies and the transformation of the structures of knowledge', *International Journal of Cultural Studies*, 10 (1): 11–20.

Miller, A. I. (2000), *Insights of Genius: Imagery and Creativity in Science and Art*, Cambridge, Mass.: The MIT Press.

Mittleton-Kelly, E. (ed.) (2003), *Complex Systems and Evolutionary Perspectives on Organisations*, London: Elsevier.

Olsen, M. (2008), 'Foucault as complexity theorist: overcoming the problems of classical philosophical analysis', *Educational Philosophy and Theory*, 40 (1): 96–117.

Paul, C. (2003), *Digital Art*, London: Thames & Hudson.

Penny, S. (1996), 'The emerging aesthetics of interactive art', *Leonardo Electronic Almanac.*
Plummer, J. (2007), *Measures of Engagement: Volume II*, Advertising Research Foundation, White Paper, March.
Qvortup, L. (2006), 'Understanding new digital media', *European Journal of Communication*, 21 (3): 345–56.
Scott, J. (ed.) (2006), *Artists-in-Labs: Processes of Inquiry*, New York: SpringerWein.
Simon, H. (1996), *The Sciences of the Artificial*, 3rd edn, Cambridge, Mass.: The MIT Press.
Sommerer, C. and Mignonneau, L. (2000), 'Modeling complex systems for interactive art', in S. Halloy and T. Williams (eds), *Applied Complexity: From Neural Nets to Managed Landscapes*, Christchurch, NZ: Institute for Crop and Food Research.
Tredennick, L. (2006), 'Post structuralism, hypertext and the World Wide Web', *Aslib*, 59 (2): 169–86.
Wilson, E. O. (1998), *Consilience: The Unity of Knowledge*, New York: Knopf.
Wilson, M. (2001), 'How should we speak about art and technology?', *Crossings: Journal of Art and Technology*, 1 (1).

8

Performance, complexity and emergent objects

Mick Wallis

1 Introduction

> Sciences must develop new methods of representing systems, new ways of seeing and interpreting what we see, and new ways of communicating new kinds of synthesis.
>
> (Jeffrey Johnson in Robertson *et al.* 2007: 285)

Two dialogues are brought to bear on the project of this book. One is between performance and design, focus of the Emergent Objects research project led by the author for Designing for the 21st Century.[1] The design of a dancing robot, SpiderCrab, addressed complexity to achieve 'performative merging' between it and human agents: the sensation that the robot was a truly improvising and responsive partner. The project as a whole was conducted according to rubrics of composition, embodiment and play; a methodology designed to set all in play to involve diverse collectives of design agents – including users – is described. The work both addresses the complex array of design determinations, and accesses intuition. The latter is the focus of the other dialogue, between performance and complexity science and theory. The role of intuition in scientific method, and scientific models of intuition, is discussed. A critique of Heidegger's writing on technology furnishes a link between questions of régimes of knowledge production – broadly the scientific and the phenomenal – and the nature of the theatrical apparatus and its cultural function.

It is suggested that, while scientific method properly pursues abstraction, this is always on the basis of more embodied understandings; and that science can gain from engaging these more methodically. Our design of an amenable future – or one at all – surely depends on both. The chapter suggests that performance has a role in such a project.

2 Theatre as complex system and as *technē*

While complexity theory was founded on specific cases in physical science, it has since been applied widely through 'analogical transfer' (Sheard 2007). One example is Hunt and Melrose's argument (2005: 78) that the 'production organism' theatre is a complex system. Drawing on Cilliers (1998), they base this identification on seven characteristics. While they principally use 'production organism' to signify any one system in particular – say, the Olivier in London – the reworking below strategically moves between specific theatres and theatre as a field of practice:

1. *Theatre as a system comprises many elements.* Consider the divisions of the space audience and performers share (stage/auditorium, stalls/balcony); the apparatuses and practices by which performers and audience bring themselves into that space and relation (box office, rehearsal); the workings of the stage (set, costume, properties, lighting, sound – the stage itself is 'syncretic' [Elam 1980], a system of systems); the ensemble of performers and other operators; the agents and systems who determine what happens (writers, publishers, directors, critics, funders, market researchers, legislators); determinations on meaning-making, pleasure-taking and group experience (social and theatrical conventions, audience composition, light/dark, commodity).

2. *The elements interact in a rich,* other-than-linear *way.* Hunt and Melrose (2005): 'the size of one element's response to a stimulus will vary according to the current state of the system, to its past state, and according to the practitioner aspiration to (citing Massumi) "qualitative transformation" '. A performer may consciously or unconsciously make their performance 'larger' in response to a 'cold' audience; other performers might follow that lead or counter it; and the audience may 'warm', be alienated or unchanged, according to their disposition as a crowd, their horizons of theatrical expectation and the prevailing physical conditions.

3. *There are feedback loops.* McCauley (1999: 246–8) identifies 'feedback loops' of 'energy exchange' between actor and audience – 'complex and subtle interchanges' present the actor with what Gielgud described as 'a never-ending test of watchfulness and flexibility'.

4. *Each element responds to local influences and cannot know the total system.* No agent or subsystem controls or knows the whole: not writer, director, star performer, funders or audience. Each is semi-autonomous. Even those presuming a panoptic view depend on locally produced understandings of the whole.

5. *Theatre is an open/co-extensive system.* Hunt and Melrose (2005) suggest that theatre meets Cilliers' criterion that 'Complex systems are usually open systems – they interact with their environment' – '(it), and its constituent elements, interact with the production organisms of other performance and the theatre industry as a whole, as well as with the wider social, cultural, political and technological environment'. But, as our reworking of point 1 attests, any one instance of theatre is simultaneously graspable as a complex system in itself (the Olivier) and as part of a wider complex system – theatre as a field of practice that embraces the public, political and economic spheres. Cilliers (1998: 4) himself states that delineation of a complex

system is often difficult, and so the convention of 'framing' is used. Rather than simply being open to outside influences, theatre and its instances can be said to be *co-extensive* with an evolving cultural context. While this can be said of any system involving human agents, it is especially pertinent to theatre. Consider Raymond Williams's (1968) seminal discussion of theatrical convention. While a dominant sense of convention then was stricture (for instance that stage speech should mimic every-day speech), Williams insists that convention is also enabling: without it, the contract of production and reception of meanings and sensations fundamental to the theatrical exchange could not operate.

Williams then develops his notion, 'structure of feeling'. Theatre and its conventions – the shape of the play, protocols for set design, acting style, audience behaviour – embody a community's lived sense of the world. It is not reducible to an ideology. While its objectified representation might be available in hindsight as 'precipitate', in the lived moment it is 'in solution'. The work of the cultural historian of theatre is to uncover and trace the correspondences and mediations between a lived culture and its embodiment in theatrical form. That understanding of theatre as co-extensive with the rest of a culture accentuates the diachronic as well as synchronic dimensions. In a later work, Williams (1977) proposes three aspects of theatrical convention: dominant, residual and emergent. Residual conventions retain force but are traces of a past structure of feeling. Emergent conventions first appear in marginal theatres or as nuances to an established repertoire, as embodiments of a present or emergent structure of feeling. While Williams uses 'emergent' in its general sense, there is more than a coincidence with its later technical use in complexity theory.

Williams suggests that theatre is a privileged place for the processing and renegotiation of cultural values. As is also classically argued for play, theatre is both part of and separate from the rest of culture: it is a place dedicated to putting cultural values and tokens into play. That play is generative variously of new understandings and dispositions, of affirmations and of clarifications.

Resisting the Marxist orthodoxy of culture as a 'superstructure' generated on an economic base, Williams models the totality as complex system. Cultural practice is itself one determination on the self-evolving whole. He calls for new methods of seeing and interpreting culture, newly understood as a particular kind of synthesis. Poststructuralist criticism in the 1980s had both positive and negative effects on Williams's project. This is not the place to discuss this, but there are suggestive correspondences between Williams's attention to culture 'in solution' and aspects of Derrida's thought, glanced at in the last section.

6. *Complex systems have history.* Hunt and Melrose (2005): 'their past is co-determining of and co-responsible for their present behaviour'. Theatre corresponds at three levels. First, within any show, how the production organism behaves at any moment is determined not only by what has been scripted but also by the way the specific performance has evolved; no two performances are the same. Second, the production organism itself evolves, as practices are embedded, transformed or rejected, intentionally or not. Third, theatre and performance as a field of practice (in,

say, the West) evolves both in its 'external' articulation with the rest of a culture, and in its 'internal' articulation. The gesture of a hand on stage co-extends both with gestures outside the stage and with an iterative history of such gestures.

The final point is the least secure with respect to theatre:

7. *Complex systems operate far from equilibrium.* Noting both 'continual change' during 'rehearsal and development' of a production, and that 'live theatre . . . is made anew each time it is performed', Hunt and Melrose (2005) add that the latter is subject to qualification: 'the stabilising strategies employed through *mise en scène*' (the particular 'staging') produce 'a state of guided, constrained change'. And, in the long-running commercial show, 'some aspects of the production organism operate in such a deterministic and stable fashion as to become non-complex'. Nevertheless, the decisive shift to 'postdramatic' performance towards the end of the twentieth century in the West – in which dialogue as inscribed in conventional dramatic texts lost dominance (Lehmann 2006) – may constitute a major self-adaptation that confirms the theatrical field of practice as being, in the longer term, far from equilibrium.

Hunt and Melrose (2005) challenge the conventionally inferior status of the 'mastercraftsperson'. Often regarded as part of the theatrical apparatus, there to serve the creative work of authors, directors and performers, the creative input of lighting operators and stage managers is either ignored or regarded as 'craft' – mapped as inferior to 'art'. This is not simply a question of regard. The dominant production organisms of Western theatre actively consign mastercraftspersons to the status of resource. Hunt and Melrose identify instances where creativity persists in such hostile circumstances.

Their argument turns on a reappraisal of Aristotelian terms – *technē, poiesis* and *epistemē* – and in particular on Heidegger's suggestion in his 1955 essay 'The question concerning technology' that modern technology (with which we may align the modern theatrical apparatus) obscures the essential connection between the *technē* of technology and the *technē* of art, or *poesis*. Both are means of 'bringing forth'. In a related argument, I have pursued a specification of the theatrical apparatus as 'a collective subject of *technē* in the Heideggarian sense' (Wallis 2005: 5). A rhetorical critique of the *Technology* essay brackets off Heidegger's reactionary romanticism, to recover more fully Heidegger's insistence on humanity's *a priori* connectedness to the human-crafted environment.[2]

Heidegger specifies *technē* as the human capacity to allow things to reveal themselves in their essence. Refusing simple conceptions of technology as 'instrumental', Heidegger asks: What is the essence of technology? He suggests that this essence is now 'in a lofty sense ambiguous'. That ambiguity is predicated on a hierarchised binary of his own construction: *technē* versus Enframing (*Gestell*). Enframing, the logic of modern technology, has not simply displaced *technē*, more discernible as the essence of technology in earlier crafts. It has also obscured *technē* as a possible essence of technology from human agents. If essence is that which endures, then Enframing now endures as the essence of technology, while *technē* endures as its prior, deeper essence (Heidegger 1977: 5, 33).

Heidegger establishes *technē* as technology's deeper essence by considering a silversmith fashioning a chalice. Medieval thought derived from Aristotle would consider the silversmith-as-maker the 'efficient cause', on an equal plane with three other causes: material (the silver), formal (chalice-ness) and telic (purpose). Heidegger deploys Greek etymology to insist that the silversmith's co-responsibility for the 'bringing forth and resting-in-itself' of the chalice rests in his allowing the other causes themselves 'presencing'. He allows them to come forth. For Heidegger, technology is 'no mere means' but 'a way of revealing'. The silversmith is engaged in a bringing-forth, or *technē*. Heidegger aligns *technē* with *poiēsis* and identifies both as bringing-forth rather than mere manufacture. Both are associated with arts of the mind and fine art as much as with craft. And both are ways to achieve *alētheia*, or truth in the sense of 'revealing and unconcealment' (Heidegger 1977: 8–13).

The revealing done by modern technology is not *poiēsis* but a challenging-forth to presence as standing-reserve. Modern technology sees the Rhine not for what it is, but as a standing reserve of energy, itself challenged-forth simply as one term in an endless chain of standing reserves. Exploited through dams, the Rhine serves up motive energy, which turbines convert to electricity, itself converted to light, which illuminates people's way to work – as mere standing reserves for exploitation in similar chains. This challenging-forth is at the behest of *Gestell*. Both modern physics and modern technology are challenged-forth by the rule of Enframing, and so insist that things bring themselves forth as calculable and as orderable in information systems. The chief characteristics of Enframing are 'regulating and securing'. It is 'the essence of modern technology' but itself 'nothing technological'. While both *technē* and Enframing are means to *alētheia*, Enframing conceals both *technē* and revealing itself from sight. And it challenges-forth the human subject as one committed to ordering, and conceiving of nature as his (*sic*) own object (Heidegger 1977: 14–25).

To be a subject, the human (individual or species) moves within and against the co-extensive world: it is constituted by the world it perceives as its object. That world includes our own cultural artefacts. The human capacity for *technē* depends on this prior articulation. Heidegger writes at this precise relation. He focuses very tightly on the silversmith to bring forth by persistent questioning the essential nature of that co-constitution. But the price is that the actual material context of the silversmith is rhetorically reduced to silence: transparent pure signals flow between silversmith, material, form and purpose. But did any real silversmith ever work under conditions conducive to *technē*? Or have all silversmiths rather also been subjects of Enframing, which annihilates *technē*? How do we begin to mediate between this abstract, idealised unhistorical figure and questions of truth, instrumentality, and not just technology but also art in the historical world? Maybe we need to pick up on *technē* less as essence and more as moment – something with both amplitude and direction – within both technology and art. And, at the self–other workface in laboratories, we might find scientists individually and as teams subtly juggling *technē* in the sense of the revealing of essences and Enframing in the sense of reduction to the measurable and calculable. Moreover, we might find moments of the casually instrumental not fully reducible to either mode of revealing.

Heidegger writes from the perspective of existentialist phenomenology. While his commitment to the specification of Being tends to the metaphysical, his figuring of the practice of *technē* as 'in-dwelling' with the other is near both to that 'tacit knowledge' observed in professionals by Schön (1983) and to what we call intuition. It is in this more mundane dimension that theatre's production of active interplay between signification and phenomenal effect (see Garner 1994: 4–16) furnishes a space for figuring our place in the human-crafted environment.

3 Science and intuition

Archimedes, Newton, Kekulé and Einstein are emblematic of the importance of intuition to scientific progress. But their *insights* perhaps have an ambiguous relationship to scientific method: are they inside or outside? Yet historiographic evidence from the 1950s to the 1980s suggests that 'the progression of scientific investigations' itself can be 'more the product of hunches and improbable connections than a series of logical and rational steps' (Fensham and Marton 1992: 115).

Fensham and Marton's ethnographic study by interview of eighty-three Nobel laureates in physics, chemistry and medicine 1976–86 identified three uses of the word 'intuition': as outcome; as process; and as individual capability. Most comments on outcome referred also to process: for instance, the sensation that there was an unidentifiable guiding hand. Many commenting on process spoke of 'subconscious process . . . a lack of a rational sense of how an idea evolved' and others of 'taste in almost the artistic sense . . . A certain rightness' (Paul Berg, Chemistry Laureate, 1980). While personal gift was claimed by few, twenty-eight identified 'knowledge or experience of the phenomenon about which intuition occurs' as important. Konrad Lorenz, Medicine Laureate, 1973, remarked:

> This apparatus . . . which intuits, has to have an enormous basis of known facts at its disposal with which to play. And it plays in a very mysterious manner, because . . . it sort of keeps all known facts afloat, waiting for them to fall in place, like a jigsaw puzzle. And if you press . . . if you try to permutate your knowledge, nothing comes of it. You must give a sort of mysterious pressure, and then rest, and suddenly BING . . . the solution comes.
>
> (Fensham and Marton 1992: 115–16, citing Marton *et al.* 1992)

In his 1968 Jayne Lectures, Director of the UK National Institute for Medical Research Peter Brian Medawar argued that 'there is nothing distinctively scientific' about the hypothetico-deductive scheme derived from Kant and refined by Popper: it is 'merely a scientific context for a much more general stratagem . . . namely *feedback*'. It accepts that 'science . . . is not *logically* propelled' but holds that 'scientific reasoning is an exploratory dialogue that can always be resolved into two . . . episodes of thought, imaginative and critical, which alternate and interact'. But what it ignores is 'the generative act in scientific enquiry, "having an idea" . . . the imaginative or logically unscripted episode in scientific thinking'. Medawar stresses that 'an imaginative or inspirational process enters into *all* scientific reasoning at every level' and provisionally maps four forms of intuition in science and mathematics: (*a*) deductive intuition: 'perceiving logical implications instantly'; (*b*) creative or inductive intuition: 'the invention of a fragment of a possible world'; (*c*) wit: 'the instant apprehension of analogy'; and (d) 'experimental flair' (Medawar 1969: 46–57). Monsay (1997: 113–16) discusses thirteen intuitive ways in which 'science gets done'.

Medawar derides the 'romantic illusion' that 'creativity [which we can understand as embracing but not coinciding with "intuition"]' is 'beyond analysis'. Its description would require 'a consortium of the talents', including 'psychologists, biologists, philosophers, computer scientist, artists, poets' (Medawar 1969: 57). Computer science in artificial intelligence has since experimented with artificial neural networks to model the generation of intuitions (Brunak and Lautrup 1990). Meanwhile, consciousness studies have pursued a more full-bodied understanding, theorising the complex interplay between cognitive, subconscious and sensory systems. Petitmengin-Peugeot attempted a 'psycho-phenomenology of intuition . . . to verify to what degree intuition is an experience which mobilizes our whole being'. Interviews revealed 'a generic structure of the intuitive experience . . . made up of an established succession of very precise interior gestures with a surprising regularity from one experience to another and from one subject to another'. The research was motivated by 'surprise at the silence surrounding the intuitive experience' (1999: 43).

In 1997, ethnographer Charles Laughlin similarly argued that, despite Bastick's description in 1982 of its isolable characteristics, intuition remained 'poorly understood and poorly studied' by psychology. While neurophysiological research demonstrates that left and right brain hemispheres perform 'analytical and integrative processing' respectively, anterior to conscious thought, Laughlin rejects assumptions that these processes correspond to 'reason' and 'intuition'. Rather, he proposes that 'ratiocination . . . refers to cognized models of cognitive processes that are culturally reified and couched in normative rules' while intuition 'refers to our experience of . . . operational cognitive processes . . . disentrained to the neural network producing consciousness'. Left- and right-hemisphere activity is common to both. Confusion arises as to the 'synthetic or analytic quality' of intuitions simply because they rise to consciousness suddenly and complete. In distancing itself from metaphysics and religion, positivist science threw 'the intuitive baby' out with 'the metaphysical bath water' (Laughlin 1997: 21–5).

A biogenetic structuralist, Laughlin suggests that the 'evolutionary advance in cognitive complexity and social cooperation' made by the time of *Homo erectus* was due to 'developments in the capacity of the hominid brain to mediate cognized relations in both space and time that were progressively less stimulus-bound'. On these foundations, language evolved specifically for 'socially adaptive purposes'. But neither it nor its 'concomitant conceptual structures' embrace the totality of 'the cognitive system and its operations'. To limit the source of scientific knowledge to language and concept is, he argues, arbitrary. This is not to exclude the need for 'socially meaningful cross-checks' (Laughlin 1997: 25–7).

Laughlin derives from Gellhorn and others a model of two complementary neuro-endocrine processes. On this basis, he formulates two linked hypotheses: that the capacity for language and concept evolved as part of the '*ergotropic* repertoire' concerned with 'consciously formulating and transmitting adaptively significant, vicarious experience among group members'; and that the capacity for intuition evolved as part the '*trophotropic* repertoire' concerned with 'developmentally reorganising the cognized environment relative to the zone of uncertainty'. Both repertoires, he argues, are appropriate to the production of scientific knowledge (Laughlin 1997: 30).

Medawar (1969: 57) suggests that, while intuition might perhaps be impossible to learn, 'it can certainly be encouraged and abetted'.

4 Intuition and emergence through embodied figuring

In Emergent Objects (EO), three subprojects each designed and prototyped one technological artefact, to achieve qualities of emergence in the interaction between designed object and human agent. The next section discusses one of these. A metaproject promoted exchange between the subprojects, setting protocols for research based on practical and theoretical dialogue between design and performance according to three paradigms: play, embodiment and composition (Bayliss *et al.* 2007 and EO website). Emergence was pursued in both *designed artefacts and systems* and *design processes*. Design visualisations are productively reductive articulations of complicated or complex fields of determinations and possibilities. Any visualisation technique is embodied: body and mind, thought and gesture are ultimately inseparable. At the December 2007 EO Colloquium, academic and practitioner experts in interaction, experience and product design, robotics, digital arts, engineering, architecture and computing participated in two workshops to pursue explicitly embodied articulations of complex design parameters.

Workshop 2 occupies territory familiar to many designers. Working to a broad design brief – to develop environments, objects or processes amenable to people with Asperger's Syndrome – small groups experimented with simple materials. The first aim was to gain close experience of the qualities of each material through play: scratching, rolling, twisting, listening, smelling. Through an iterative and sometimes enfolded process of playful experiment, composition and evaluation, each group developed an installation for discussion by the whole group. Each material stood both for itself and for unspecified other materials; and each installation occupied its actual space while indicating other possible scales or extensions. This circulation between concrete and metaphoric, based in close bodily engagement, was enabled by Workshop 1, which used techniques not yet widely practised in design.

Workshop 1 combines two phases. The Embodiment Phase derives from Brazilian theatre practitioner Augusto Boal, whose work is fundamentally concerned with empowerment. The work is participatory. There is not spectator and actor: each human agent as 'spectactor' rehearses in dramatic action the means by which they might come to act in everyday life. Boal's techniques situate the spectactor's body in a composed and dynamic space, where the relations between bodies, their individual and collective gestures and actions comprise a collective and transactional image of the experienced real. They correspond somewhat to what Paul Crowther has termed a 'sensuous manifold': an 'integral fusion of the sensuous and the conceptual which enables art to express something of the depth and richness of body-hold in a way which eludes modes of abstract thought' (Crowther 1993: 5). Working in both semiotic and phenomenal registers, the embodied composition constitutes an emergent projection of the world as experienced by those who have created it. Crowther's concern is with the painter's expressive contribution to the viewer's grasp on the world. Boal's and ours is with the collective composition of sensuous manifolds by constituencies for their own use. The composition does not comprise clear signs: it tends towards Kristeva's 'semiosis' – *potential* signifiers in a state of flux. This and the phenomenal dimension generated by body work generate that Heideggarian 'knowing' (as opposed to objectified knowledge) foregrounded by Crowther and informing the EO project.

The Projection Phase derives from Isabel Jones of Salamanda Tandem, developed by her and Wallis as pedagogic tool in Arts Work With People (AWP). A central strand of Jones's work is participatory arts experience for people with severe access needs. Participants are assisted in negotiating the best form of artistic expression for their needs and desires. While a single art form might be delivered (in, say, a day centre for people with learning difficulties or a drop-in dance group for people with a visual disability), an important technique is playful progression from one matter of expression to another in a single session: say, from making vocal sounds to moving the body to making marks. Psychic material or aesthetic impulse is carried from one medium to the next, according with Prestinenza Puglisi's (1999) expansive use of 'projection' to embrace both architectural drawings as two-dimensional projections of three-dimensional space, and a series of artworks, each a response to the last. The material is 'thrown' from one system into another.

Boal's Forum Theatre might be called a dynamic embodied visualisation generated by an oppressed community of their always complex situation, used by them with expert assistance to rehearse both their situation and possible modes of resistance.

His Image Theatre – the basis of the Embodiment Phase – focuses on 'internal' oppressions. Boalian practice is pertinent not only to design visualisations but also to scientific modelling of complex systems. The EO workshop is not offered as an alternative to either, but as a valuable supplement, in Derrida's sense of an addition that changes that to which it is added – for two reasons. First, computer models and design visualisations both derive from observation and induction: that which is to be modelled or visualised is selected and defined, and usually variables with concrete correlates are identified and quantified. Boal-derived practice brings users and/or those who are themselves part of the complex system fully into the conceptualisation of the model itself. It is more than a consultation tool. The field of sustainability science is a good example of its potential benefit. Second, while design visualisations are usually static or at least non-complex, Boalian techniques offer a modelling tool that is both dynamic and capable of precipitating out clear, diagrammatic images from an iterative process of complex manipulations. And, while computer models can tend towards silos of mathematical abstraction, Boalian techniques could act as points of iterative dialogue between modelling outputs and correlates in social reality. Jones's work correlates with Boalian techniques in its emphasis on participation and empowerment through creative figuring. It also offers a valuable extension to a third aspect of Image work, the transformability and malleability of the image, figure or sign – both as a substance and as received and understood or felt by participants. This can be best described through a description of Workshop 1 (Figure 8.1).

Workshop 1 works best with a group of about sixteen plus Facilitator(s). A simple Boalian exercise inducts the group: After physical warm-up, the group forms pairs. A and B shake hands and freeze. B steps away and rejoins the image with a different composition of their own body – say, kneeling down with their neck in contact with A's outstretched hand, as if being strangled. Freeze. A steps away and rejoins – and so on. The facilitator insists on spontaneity: no conscious thought. The group are quickly brought into a state of exchange, embodied response and free composition. As play, the work gives participants subtle access to psychically cathected mind–body formations in an unthreatening way.

Figure 8.1 Co-pilot exercise at EO Colloquium

The work moves to Boal's Co-pilot exercise (Boal 1992: 164–200): Pairs relax one another to the floor, and each partner finds their own space. Eyes closed, they think about a 'blockage' important to them. *Try to picture it. How does it feel? Who's involved? How would you describe it?* Groups usually consist of individuals who want to address a problem that is common to them – like alienation at work. In this case, feeling blocked or disabled by the built environment had been raised as the topic of the workshop a few hours before. The pairs re-form, and A as Pilot recounts their 'scene' to B as Co-Pilot. *Don't explain or analyse. Just tell.* B listens and checks details until they feel they could make the description themselves. They swap roles, and the process repeats.

Next, one pair offers two scenes to the rest of the group. B leaves and A 'bodysculpts' as many people as they want into a representation of their blockage. *The bodies can be people, things or forces – like a barrier to communication. Is there tension in the bodies? Make sure they are in exactly the right place and posture. Don't explain or analyse. Just sculpt. Include yourself in the Image.* Once A is satisfied, all make sure they understand the full Image and their embodiment within it. Then A leaves the room, and B as Co-pilot returns and goes through the same process, sculpting A's blockage with no knowledge of what A has sculpted.

The group reassembles, and both A and B's Images of A's blockage are repeated in turn. *How can we describe this? Don't explain. I see bodies scattered, all stooping except one on the margin reaching skyward . . .* Sometimes there are close similarities between the two Images. But often they vary either in some significant detail or in their whole shape. *The Co-Pilot's Image is both a gift and a potential insult. It embodies the Pilot's situation from within but through different eyes.* Next, A and B take turns in sculpting the Anti-image of the two Images they have made, starting from the Image and working as if unravelling it, moving bodies towards an image in which the blockage is undone.

There is rarely time to pursue all potential Images in the room. But, as the Facilitator offers, not explaining or narrativising makes any Image a metaphor available to each on their own terms. Boal refers to this as '*metaxis*' or 'the state of belonging completely and simultaneously to two different autonomous worlds' (Boal 1995: 43). Especially for 'non-performers', moving the body within a psychically cathected but safely distanced metaphor can have epiphanic effects. Long-lived emotional knots can 'undo themselves'. While this is not the direct aim in the context of a design process, the fact that the exercise can engage a group in embodied imagination that accesses deep experience makes it a powerful preparation for Workshop 2. But, as argued above, it has its own immediate pertinence for designers and those that employ them, either as stand-alone or in combination with the Projection Phase.

Consider artist Julian Burton's visualisations of corporate systems deployed by Eve Mittleton-Kelly (Art in the Science of Complex Systems website). Where workers feel that the system is blocked is made apparent. But the composition is ultimately that of the artist – and tends sometimes to a *perspectival* view. Co-Pilot generates compositions that *embody from within* both a structure and what it feels to be part of it.

Its rehearsal prepares individuals and groups to move the structure; or can provide direct testimony to managers. When addressing corporate systems design, it foregrounds how particular design processes and agents are situated within corporations. The application of Boalian techniques in a wide range of contexts including education and management is well documented (e.g. Schutzman and Cohen-Cruz 1994). Wallis with others has used versions of Workshop 1 with care workers, university lecturers, health workers, scientists, property managers – and urban planners: *Where did they feel blocked in relation to their desire to serve a community?*

In Boal's Co-Pilot proper, nobody writes. In Workshop 1, the group records key words and phrases offered to describe the Images, for iteration in the Projection Phase. The words and phrases are pooled on the floor and each participant selects four. Subgroups of around five pool their material. The word-text is developed by individuals or pairs into rants, short poems, lists, word-maps, dialogue, sermons, slogans. The subgroup collectively arranges a selection of these word-texts into an array, using both floor and wall; the logic does not matter – the act of collective composition (selection, combination) does. They project some by OHP. *Speak, sing, chant, or whatever the texts singly, in pairs or groups. Develop sequences, routines if you like.* (After five minutes) *Play with making body sculptures in response to the texts and the vocalisations you've developed.* They then draw or make other marks on paper or acetate and develop three-dimensional structures from any of the simple materials in the room in response to what they have made so far. Finally, they compose an agreed selection of their materials into a performance installation – something that combines sound, words, movements and images. *Include your voice work.* They rehearse and refine until ready to show.

When all are ready, the workshop moves to making sound circles. Each group forms a circle, arms resting on partners' shoulders. *Relax, breathe in through nose and out through mouth. Imagine your breath reaching down the well. Introduce some vocalisation. Send it down the well. As you vocalise, listen to the sounds in your circle. Play with the sound you're making: pitch, volume, shape, length. Play with plosives, linguals, drones . . . Work your voices together as a group . . . Rest. Start again with drones and develop something that has a beginning, a middle and an end. No rehearsal. Just do it now.*

Each subgroup visits the installation of another subgroup and vocalises in response, using the sound ensemble technique they have developed in the sound circle. *No need for a circle. Perform <u>to</u> the installation.*

The Projection Phase is designed to continue the work of *metaxis*, now projecting the field of mobile but cathected metaphors though different matters of expression, each with its own mode of engagement with the body (spatial, tactile, social, solo, carried by the diaphragm . . .). It stimulates lateral thinking, the imagination. Meanwhile, the psychic engagement remains. It is something like a waking dreamwork – Freud draws attention to the work done in dreaming as opposed to any supposed latent content (see Laplanche and Pontalis 1973).

In the journey to the sensuous manifolds constituted by the collectively performed installations, Workshop 1 does three kinds of work in relation to the design process: *attitudinal work*, opening designers and those designed for alike to imaginative exploration; *processing work*, engaging with embodied experience and its psychic counterpart; and *preparatory work* towards Workshop 2. While less precise than many visualisations, the images generated do nevertheless *map* significant factors in relation to one another. In doing so, they can usefully address user experience. Where accessed by design agents, this has the quality of empathetic (feeling with) rather than sympathetic (feeling for) or objectified engagement. Through the operation of *semiosis*, *metaxis* and *embodied knowing*, the images have a generative mobility when witnessed, stimulating lateral speculation or intuitive leaps – by designers and the designed-for. The Workshop both addresses complexity and insinuates a productive complexity into the design process.

5 SpiderCrab and the embodied conversation

We encounter two related forms of resistance to Emergent Objects methodology from a small minority, typically physical or social scientists. One is discomfort with the shifting of 'notation': 'I'm comfortable with number as metaphor. Moving from system to system was deeply uncomfortable – though interestingly.' Many scientists knowingly engage metaphorical speculation. But it does indicate a professional *habitus* – defined by Bourdieu (1998) as 'the deeply-installed set of cultural frames within which our physical improvisations can occur' – to which the facilitator needs to be sensitive. Medawar (1969: 58) describes the scientist too exclusively wedded to Popperian falsification: 'often unproductive, as if he had scared himself out of his wits'. The other is a twofold performance anxiety: fear of exposure and feeling threatened by being pressed for an outcome. There is a delicate balance to be struck in order to nudge and not harass, to manage stress. *Habitus* suggests a circulation between bodily composure, daily practice and mental universe. Both bodily exercises and cycles of play and reflection were deployed by EO – and SpiderCrab in particular – to nudge *habitus*.

SpiderCrab was conceived as a multisensory 3.5-metre-high six-legged robotic cross between architectural environment and dancing partner. One limb has been constructed and evaluated with sensorium limited to vision. McKinney *et al.* (2008) discuss design process and solutions; the concern here is with the latter, in particular software architecture. The aim of the transdisciplinary team was to achieve 'performative merging' between robot and human partner. Adapting the Turing Test, we conceived an 'embodiment test' with the criterion that the dancer should feel they were dancing with a 'true' partner.

In dialogue with the team, John Bryden developed an '*interlingua* for robotic dance'. SpiderCrab detects its human partner's movement through a vision system tracking a green arm-band. The robot's movements are generated from this data. The robot both 'sees' the human partner and itself then moves on the basis of the same language – Laban Movement Analysis.

Laban focuses on quality of movement rather than on pose or position: light/strong, direct/indirect, free/bound, sudden/sustained. Translated from models of play as framed activity, the software architecture has the simple form of four successive levels$_L$. The *interlingua* operates as a bias$_2$ on the foundation of random generation$_1$ of the robot's movement. The robot adopts the basic modes$_3$ dancers use in improvisatory duets – Copy, Oppose (e.g. light in response to strong), or Innovate. (It may also Follow the dancer's position in the room.) The modes can be programmed to vary in sequence, duration and combination. We call this disposition$_4$. SpiderCrab might be disposed to Copy for a while and then Oppose.

Evaluations identified a sense of an *offer* coming from SpiderCrab whereby its gesture or sequence calls forth a response; its own *response* to one's own movements, while not being slavishly bound to them; its nature as *embodied agent* – the sense that it appears to have a historically achieved *habitus*; and *friendliness* – compatibility with human agents in terms of general quality of movement, behaviour and physical being.

SpiderCrab is engineered to make a persistent 'offer' sufficient for people to feel that they are dancing with a 'partner' and so enter into a contract of mutuality – performative merging. Dancer and robot engage in a dance improvisation, an embodied conversation. Conversations are self-generating and unpredictable. The SpiderCrab–dancer couple is, we argue, a complex system. It has the quality of emergence. The software *approaches* complexity by means of its levels. It may be considered '*quasi*-complex': it lays the foundation for complex interaction, and may appear to the human dancer to be complex, as she herself is.

It might be objected that the robot–human couple is not complex but merely complicated on two grounds: that it is in fact closed; and that it is not self-evolving as a system. The first might be tempered by the principle of co-extension characteristic of human performance; and the second by framing emergence to the dance interaction – though, in the wider frame, we have recorded changes in dancers' repertoires of improvisation – their 'signature' or dancerly *habitus* – after engagement with SpiderCrab.

While developed for a specific artefact and deploying a specific categorising system (Laban), the *quasi*-complex system achieved in SpiderCrab offers a transferable model for 4D design (Robertson *et al.* 2007: 286–7) in pursuit of performative merging.

6 Ritual and ritualisation

While their ability is seldom paired with a facility for mathematical concepts (Treffert 2006: 252), several people with Asperger's Syndrome – like Daniel Tammet (2006) – can make large mathematical calculations at high speed, often experiencing number sensually – perhaps as colour. They also find the built environment and everyday social exchange too 'noisy': it stresses them. Shamans have been called 'technologists of intuition' for their ability to enter at will into meditative states through ritual trance. Laughlin

(1997: 30) reports studies of such access to alternative consciousness in terms of 'ergotropic–trophotropic tuning'. With regard to the trophotropic, he notes that both play and playing around with ideas are 'manifestations of the intrinsic . . . cognitive imperative that drives (the construction of) adaptive meaning'. The 'individual cognitive system' has been observed to work best under positive stress (following Selye, '*eustress*') and worst under negative stress ('*distress*') (Laughlin 1997: 29). Workshop 1 tunes between these types of stress. But are there models recommendable to scientists as a ritual route to intuition?

The mission of the Theatre Research Workshop (TRW), founded by Nicolas Nuñez in Mexico in 1975, has been

> to safely and effectively transplant religious rituals into contemporary secular cultural forms, maintaining their psychophysical and numinous potential beyond the matrices of their belief-system of origin.
>
> (Middleton 2001: 45)

Participants achieve extra-daily experience through 'structured sequences of psychophysical actions' – 'dynamics'. The ultimate aim is 'anthropocosmic theatre' – a means to return to the human organism 'its capacity to be the echo box of the cosmos'. '(M)ental, physical, and energetic discipline 'deprogram[s] . . . daily behaviour patterns' to reach 'an experience of intense presence within the moment' (Middleton 2001: 43–7, citing Nuñez).

TRW pursues a non-religious route to the dimensions of our existence that modernity historically but perhaps temporarily wrested from us: the 'secular sacred'. Now, as well as physical actions, each dynamic has an 'an internal score' – meditation focused on images from pre-Hispanic mythology. Nuñez insists that these figures and narratives are purely 'a focusing device'. Their multiple meanings allow us 'to experience dualities in such a way as to undermine apparent contradictions in the face of meaningfulness on an a-rational level' (Middleton 2001: 46, 62). But there are echoes perhaps of Eliade's 'hierophany' – the belief that 'mythic sensibility' resonates with a pre-Edenic timelessness, ritual 'invok(ing) the moment in which order was created out of chaos' (Sheard 2007: 200). Even so, TRW might point to a fully agnostic means to 'activate our own latent psychophysical resources' through ritual. This might supplement the more daily-embodied practices of Boal.

Ritual here points in two directions. For Nuñez, it is a psychophysical enabler. Yet experience of the 'numinous' in other contexts is actively connected to religious or mystical totems through what Catherine Bell (1997) calls 'ritualisation': the layering of unconnected oppositions (high/low; inner/outer; action/thought . . .) to suggest that the terms are aligned and subtend to some grounding or transcendent principle. Wallis (2005) identifies ritualisation as the rhetorical form of Heidegger's 'Technology' essay. Medawar (1969: 44) implicitly invokes the model in his critique of inductivist science as 'mumbo-jumbo', and Lyotardians might find it underpinning Laughlin's cultural reification, ratiocination. It also maintains the restricted *habitus* of Medawar's ultra-Popperian scientist. Any invocation of ritual in a methodology for addressing complexity would need to take care about ritualisation.

Sheard (2007) questions the extension of complexity theory beyond specific instances of physical science. Is such metaphoric extension rigorous? 'How deep does metaphor go?' he asks. 'Are not the metaphors of complexity merely decorative?' (p. 195). Yet at 'a primary level of scientific . . . analysis' a concept like 'emergence' is itself 'coloured with metaphor as it is conceived' (p. 206). Conceptualisation and theorisation bring metaphor in train. So it matters how metaphor can both be valued as a scientific tool and itself be opened up to enquiry in terms amenable to science. Sheard identifies two tensions among others in attempts to synthesise phenomenology and complexity theory. One is between phenomenology's pre-epistemic basis and scientific principles (pp. 187–8). The other is between conceptions of metaphor that attribute to it the ability to access ontological 'Being' or cosmic 'Truth' and those that do not. For the argument of this chapter, Lakoff and Johnson's notion, derived from the non-metaphysical phenomenology of Merleau-Ponty, of metaphor as 'a kind of "imaginative rationality" ' rooted in the circumstance that 'we are coupled to the world through our embodied reactions' (pp. 195–6), provides a materialist link to, for instance, Laughlin's anthropological neuropsychology.

If, following Medawar, we are interested in understanding the nature of generative thought rather than – if we followed Heidegger – the nature of 'Being', then attempts to link Derrida's *différance* to neural function as a contribution to cognitive science is suggestive – even if it must remain at the level of analogy. Derrida's neologism combines the notions, first, that signification is always predicated on *difference* (green is not red) and, second, that ultimate meaning is always *deferred* (what precisely is green?). If meaning emerges from the play of signifiers, then this is always as a fleeting trace, as in writing. Changing focus, we might turn again to Workshop 1, where the embodied emergent metaphors arguably constitute a Derridean trace, less fixed than writing and mediating more fully between *ergotropic* and *trophotropic* consciousness.

The mission here has not been to claim ascendancy of either intuition or phenomenology over scientific understanding. Rather, the residual presumption in scientific discourse of superiority over and freedom from intuition and the metaphoric is resisted. While scientific method properly pursues abstraction, this is always on the basis of more embodied understandings. Science can gain from engaging these more methodically; and the biological and computational basis of both metaphor and intuition are proper targets for scientific understanding. Our design of an amenable future – or one at all – surely depends on both. This chapter suggests that performance has a role in such a project.

Notes

1 'Emergent Objects: Designing the human/technology interface through performance' funded by AHRC and EPSRC: Grant Reference AH/E507522/1.

2 The next five paragraphs closely follow Wallis (2005).

References

Bayliss, A., McKinney, J., Popat, S. and Wallis, M. (2007), 'Emergent objects: designing through performance', *International Journal of Performance Arts and Digital Media*, 3 (2–3): 269–79.

Bell, C. (1997), *Ritual: Perspectives and Dimensions*, Oxford: Oxford University Press.

Boal, A. (1992), *Games for Actors and Non-Actors*, trans. A. Jackson, London: Routledge.

Boal, A. (1995), *The Rainbow of Desire: The Boal Method of Theatre and Therapy*, trans. A. Jackson, London/New York, Routledge.

Bourdieu, P. (1998), *Outline of a Theory of Practice*, Cambridge: Cambridge University Press.

Brunak, S. and Lautrup, B. (1990), *Neural Networks: Computers with Intuition*, Singapore/London: World Scientific.

Cilliers, P. (1998), *Complexity and Postmodernism: Understanding Complex Systems*, London: Routledge.

Crowther, P. (1993), *Art and Embodiment: From Aesthetics to Self-Consciousness*, Oxford: Clarendon Press.

Elam, K. (1980), *The Semiotics of Theatre and Drama*, London: Methuen.

Fensham, P. and Marton, F. (1992), 'What has happened to intuition in science education?', *Research in Science Education*, 22: 114–22.

Garner, S. B. Jnr (1994), *Bodied Spaces: Phenomenology and Performance in Contemporary Drama*, Ithaca, NY: Cornell University Press.

Heidegger, M. (1977), 'The question concerning technology', in *The Question Concerning Technology and Other Essays*, trans. William Lovitt. New York: Harper & Row, pp. 3–35.

Hunt, N. and Melrose, S. (2005), 'Technē, technology, technician: the creative practices of the mastercraftsperson', *Performance Research*, 10(4): 70–82.

Laplanche, J. and Pontalis, J.-B. (1973), *The Language of Psychoanalysis*, trans. Donald Nicholson-Smith, London: Hogarth Press.

Laughlin, C. (1997), 'The nature of intuition: a neurophysiological approach', in R. Davis-Floyd and P. Arvidson (eds) (1997), *Intuition: The Inside Story. Interdisciplinary Perspectives*, New York/London: Routledge.

Lehmann, H.-T. (2006), *Postdramatic Theatre*, trans. with introduction by Karen Jürs-Munby, London: Routledge.

McCauley, G. (1999), *Space in Performance: Making Meaning in the Theatre*, Ann Arbor, Mich.: University of Michigan Press.

McKinney, J., Wallis, M., Popat, S., Bryden, J. and Hogg, D. (2008), 'Embodied conversations: performance and the design of a robotic dancing partner' in *Undisciplined! Proceedings of the 2008 Design Research Society Conference*, Sheffield, 16–19 July.

Medawar, P. B. (1969), *Induction and Intuition in Scientific Thought*, London: Methuen.

Middleton, D. K. (2001), 'At play in the cosmos: the theatre and ritual of Nicolas Nuñez', *TDR*, 45 (4): 42–63.

Monsay, E. M. (1997), 'Intuition in the development of scientific theory and practice', in R. Davis-Floyd and P. Arvidson (eds) (1997), *Intuition: The Inside Story. Interdisciplinary Perspectives*, New York/London: Routledge.

Petitmengin-Peugeot, C. (1999), 'The intuitive experience', *Journal of Consciousness Studies*, 6 (2–3): 43–77.

Prestinenza Puglisi, L. (1999), *Hyper Architecture: Spaces in the Electronic Age*, trans. Lucinda Byatt, Basel/Boston, Mass./Berlin: Birkhäuser.

Robertson, A., Lycouris, S. and Johnson, J. (2007), 'An approach to the design of interactive environments, with reference to choreography, architecture, the science of complex systems and 4D design', *International Journal of Performance Arts and Digital Media*, 3 (2–3): 281–94.

Schön, D. A. (1983), *The Reflective Practitioner: How Professionals Think in Action*, New York: Basic Books.

Schutzman, M. and Cohen-Cruz, J. (eds) (1994), *Playing Boal: Theatre, Therapy, Activism*, London: Routledge.

Sheard, S. (2007), 'Complexity theory and continental philosophy Part 2: A hemeneutical theory of complexity', in P. Cilliers (ed.), *Thinking Complexity*, Reading: ISCE, pp. 187–210.

Tammet, D. (2006), *Born on a Blue Day*, London: Hodder & Stoughton.

Treffert, D. A. (2006), *Extraordinary People: Understanding Savant Syndrome*, revised edn, Lincoln, Nebr.: iUniverse.

Wallis, M. (2005), 'Thinking through *technē*', *Performance Research*, 10 (4): 1–8.

Williams, R. (1968), *Drama from Ibsen to Brecht*, London: Chatto & Windus.

Williams, R. (1977), *Marxism and Literature*, Oxford: Oxford University Press.

Websites

Arts Work With People Project, http://www.leeds.ac.uk/paci/artworkpeople/purpose.htm

Art in the Science of Complex Systems (ECiD), http://www.complexityanddesign.net/art&complexity.html

Emergent Objects, http://www.emergentobjects.co.uk

9

Developments in service design thinking and practice

Robert Young

1 Introduction

In 2002, Nicola Morelli wrote: 'In the design discipline the methodological implications of product/service systems rarely have been discussed even though design components play a critical role in the development of these systems' (Morelli 2002). Criticism of design practice domains lacking a theoretical framework and methodological basis is typical of the theory–practice nexus in design. The overview of the work of the 'Embracing Complexity in Design' community commented that: 'It is now recognised that the objects of most domains are "complex systems", including natural systems, social systems and artificial systems. The science of complexity cuts across the particular domains seeking principles and methods that can be applied to complex systems in general.' Johnson *et al.* 2007).

This chapter looks at the evolution of the theory–practice nexus for service design as an area of complex design practice, over the relatively brief period of its emergence. It takes its working structure from the elements of an integrated model of designing to aid understanding of the complexity paradigm in design practice, which has been developed by the author as part of his work contributing to the 'Embracing Complexity in Design' community (ibid.). These elements refer to the content, context and process of designing, where the underlying purpose is to conceptualise and visualise the content and context of complex designing, rather than concentrating exclusively on the rationalisation of its process (Young 2008a). The main aim of this work is to concentrate on the role of method in support of design practice – more specifically, identification of the theory that most aptly reflects the evolution of service design practice.

While the work can be considered more *a posteriori* in its nature because the theoretical framework to describe and support the practice of service designing tends to be evolving dependent on the experience, which is again typical of the development of the theory–practice nexus in design domains, in contrast there are aspects of the evolution of theory to support service design that might be considered *a priori* in nature because they appear at first to be independent of the experience. This chapter puts forward new thinking as its contribution to supporting the evolving theory of practice for service design that is based on the former but configured by the latter approaches. Perhaps an analogy can be drawn here with Chris Jones's thinking over thirty years ago when he said: 'I sometimes think of designing as a meta-process, occurring before the product exists, that can predict enough of the future to ensure that the design can have the same quality of rightness that we see in natural organisms that have evolved naturally, without design.' The author has previously questioned whether this analogy is appropriate for the design of services, as opposed to products and artefacts (Young 2008b). The significance here is that Jones was perhaps one of the first design researchers to model the different levels of complexity of design beyond components and products, to encompass systems, services and communities. He also exhorted designers to be more aware of the impact of their work on society at a systemic level. Jones identified the need to appreciate products by understanding their whole. He talked about the social, economic and political basis of the existence of a single product in order to address human needs. He recognised a need for change as a result of the increased complexity of new products brought about by technological developments.

In his reassessment of the nature and utility of design methods in the late 1970s, Jones asked what design methods were and proposed that they were 'anything one does while designing . . . any action whatever that the designer(s) may decide is appropriate', emphasising the relevance and utility of methods to the practice of design (Jones 1992). In his consideration of the usefulness or purpose of design methods and processes, Jones wanted to provide an adequate way of 'listening to the users and to the world in such a way that the new design becomes well fitted to people and to circumstances' (Jones 1977). This is a laudable objective given the nature of service design, as determined through the early conferences on the subject, described in this chapter.

Returning to the structure of the chapter, the explanation of shared thinking about service design methods is by reference to the *context* in which this has evolved: the various conferences and workshops on and around the subject. Service design practitioners and researchers have reflected in their presentations on the experience of their practice and how this has given rise to the refinement of design *processes* and understanding about their utility in different contexts of complex service design problems for private-sector organisations, public-sector organisations and engagement with social communities. The conclusion to this work is derived from a discussion of the theoretical *content* issues that have both preceded and succeeded service design experiences, particularly how the disciplines of design theory, complexity science and social science theory can inform and support the practice of service design. The final conclusion from the 'Designing for Services' workshop was that service design can be viewed through a number of different disciplinary lenses, not least those mentioned above, because it is fruitful to conceive of it as an emerging interdiscipline ('Designing for Services' was another Designing for the 21st Century research project, led by Lucy Kimbell and Victor Seidel from the University of Oxford, Said Business School: Kimbell and Seidel 2008).

2 The evolution of service design thinking

Key developments and understanding that have contributed to awareness, understanding and the evolution of service design thinking have come from a variety of conferences and workshops around the Western world. These conferences have dealt with basic questions such as the what, where, when and how of service design. They looked at the experience of designing for complex product/service and service/systems, predominantly in business-sector contexts, and the types of methods, processes and techniques that are being used to undertake this type of work. The most recent workshops and symposia have begun to look more closely at theoretical and design research issues in detail, particularly in the context of designing in the public domain. Critical observations and key aspects emerging from these conferences are referred to here in a chronological order.

ISDn and ISDn2

International Service Design Northumbria (ISDn) was the first dedicated conference on the subject of service design. It took place at the Sage Music Centre, Gateshead, in the United Kingdom in March 2006 ISDn 2006. Since then, the ISDn series of events has formed a platform for exploring the emerging field of service design. ISDn looked at service designers – who they are and what they are doing. ISDn2 followed in November 2006, and explored the relationship(s) that service design as a design subdiscipline or interdiscipline might have with business.

These two conferences addressed a broader purpose concerning the nature of design practice to support the role of design in society. A focal interest here is the development of appropriate descriptions of service design *content* and *process* to improve the designer's ability to navigate and contend with complex projects, including the relationship to people as subjects and objects of research. The approach was the perspective of the design practitioner and their sense-making requirement for theory and knowledge to support design practice. The theoretical interest lies in the epistemology of design practice – including its relationship to real-world problems, to other disciplines, their philosophy and methods. The keynote content in ISDn included:

1. Descriptions of service innovation through design thinking; *Prof. Tim Brown, IDEO US*
2. Social, technical, economic and environmental signposts for the next decade; *Dr Andrea Cooper, Design Council*
3. Pioneering service design practices; *Chris Downs, Live|Work*
4. Corporate business perspectives on the objects of service; *Prof. Steven Kyffin, Philips Design*
5. Designing design in a complex world; *Prof. Bob Young, School of Design, Northumbria University*
6. Better services happier customers; *Oliver King, the Engine Group*
7. Redesigning public services; *Jennie Winhall, RED at the Design Council*
8. Conclusions from an ethnographic study of the cultural traits of professional design practitioners by a business management researcher, 'Designers! Who Do You Think You Are?'; *Kamil Michlewski of Northumbria University*

ISDn set the scene for similar design conferences elsewhere on the topic 'What is service design?' Emerging issues were:

1. Designing with people not for people
2. Designing with multidisciplined teams
3. Learning to listen before acting
4. Service design needs a sophisticated multidimensional model of content and context issues not just design process elements
5. Design students are becoming more concerned about issues to be addressed by design rather than just learning the skills of a traditional design discipline
6. If product design requires a developed sense of seeing manifested through the act of sketching, then service design requires a developed sense of listening manifested through the act of storytelling.

ISDn2 was set up as a workshop supported by the 'Embracing Complexity in Design' project. The workshop deliberately looked at the context of service design and business. After keynote presentations by Hollins, Van Halen and Young ISDn2 2006, it posed questions for its audience to resolve:

1. How can services be prototyped (both demand- and supply-side perspectives)?
2. What kind of data is useful to the designers of a service and other stakeholders in the service, during the design process?
3. What criteria are used to define a service as successfully designed (i.e. profit-turning, easy to use and quality of user experiences, ecologically sustainable etc.)?
4. What methods – visual or verbal or conceptual – are available to help articulate complex systems?

The workshop plenary involved a dialogue between speakers and members of the audience to address further questions that had emerged, including:

1. In what ways do the designs of public- and private-sector services differ?
2. What influences the perception of risk amongst public- and private-sector service design sponsors?

Emergence Conferences and the Design for Services Workshop

Much of the content of presentations made by service design practitioners at the Sind conference was repeated at the Carnegie Melon–Emergence Conference in September 2006, together with interesting and useful North American practitioner perspectives. The second Emergence conference in September 2007 aired practitioner reflections and ideas about service design that also appeared in the year-long workshop 'Design for Services'. The purpose of Emergence 2007 was to look at the boundaries of service design, developing the earlier theme: what service design is about. It included a presentation by Jennifer Leonard, where the debate that followed was particularly helpful in revealing the boundary issues on the subject (Leonard 2008). She observed that the quality of service depends on both the responsibility and the accountability of the deliverer and the receiver, on our ability to give to others as well as on our capacity to receive. She maintained that service deliverers need to understand receivers' needs, that every single service experience depends on human encounters, also on the power of stories. It is about dialogue between people, interconnectivity, trusting others, and surprise. Leonard stressed her view that service design is not a subset of product design but a meta-discipline, like a universal contract between clients and designers. The audience at that conference felt this to be an audacious concept. Can design do this? While other disciplines have attempted this in the past – i.e. participation design in architecture – is service design something that can be accepted by other disciplines as the new 'universal joint'? Not just design disciplines but law, engineering, business and marketing etc. How can this possibly happen? Leonard's view is that this level of activism in not just about a new or evolved discipline but a way of life!

Both the ISDn and Emergence conferences and the 'Design for Services' workshop referred to the importance of using stories to engage others. Personal experiences are powerful in making businesses listen, to promote interest and open-mindedness in the business context. Design and innovation discussions in *Business Week* show that more and more businesses are thinking like designers (Leonard 2008); that this also has an ecological dimension. Business consideration is not just about product and design economies but also about ecologies. This aspect was brought home to the audience of one of the 'Design for Service' workshops in a presentation by Jim Spoher, founder of the IBM Service Science Almaden Research Center in San Jose, when he explained that IBM had moved beyond the strategy of a limited range of heavily marketed, branded product lines to operate increasingly as a viral business built on a myriad of product-service ecologies that contribute to and collaborate with many other organisations. Spoher maintained that this creates a much more resilient business model because it is less prone to downturns in any particular market area. The underlying concept to this sounds very similar to the argument put forward by George Rzevski in the 'Embracing Complexity in Design' workshops that 'more is more', rather than design's traditional mantra of less is more; that society and business need to understand that the future is about 'multi-agent systems', where each agent is autonomous and interacts with others in its neighbourhood using 'local' rules that emerge based on behaviour learned from interactions, thus creating viral connections to establish new opportunities, advantages and benefits (Rzevski *et al.* 2006): in other words, a truly complex, intelligent, self-regulating and evolving operating system, where the inherent complexity pre-empts conventional methods of analysis. Of course this is the *bête noire* of many science fiction stories such as *Terminator*, *I-Robot* etc., where the technology begins to perceive that it is the human element that is the weakest link in the system – because humanity is the very thing that has not been accommodated in a technologically determined complex system.

Service design conferences, and events to date generally, conclude that a lot is lost in how we interpret design in complex systems. The recurring question is how do we bring people into systems? There is common agreement that we can so easily lose qualities of human interaction. Systems need to look from a human-centric not a technology-centric or technologically determined perspective. The holy grail of service design is how we bring humanity and compassion and empathy into communion with technology. 'If people have empathy with a system then this makes the experience of the system a lot more memorable, pleasurable and rewarding' (Leonard 2008). Experience is often more about people that you encounter than about the system or things in a service. Good experiences implant memories that are powerful, which has a greater influence than brand.

In his closing remarks to Emergence 2007, Dick Buchanan referred to the critical difference between a system designer and a service designer. How to interface and connect with problems that have many interactions? Also, what is the relationship between service design and management? Is good service design just good management, i.e. workflow and people management? Buchanan stated that management is a kind of design activity, but what do designers do that is typically different from managers (Buchanan 2008)? Similarly, the 'Design for Services' workshop had tried to resolve the essential difference between the work of a service designer and a business strategy consultant.

Buchanan summarised by saying that artefacts can be portals to the services offered by companies. Designers have the ability to see the whole as well as parts of the problem. Managers seem to have lost this in management practice! He pointed out that another big difference is the ability of designers to visualise and their skill of embodiment – to make real. He added that we are still physical and emotional beings, and that service design is not a bloodless process. He remembered that, back in the 1960s, Herbert Simon had commented that engineering had lost touch with people and that perhaps management has done the same!

Developments from the Dott 07 programme

Concurrent with the first service design conferences and events, the Designs of the Time 2007 (Dott 07 2007) programme took place and was the source of much experiential knowledge and ideas concerning designing with people through community engagement. Dott 07 was the first of a ten-year programme of biennial events developed by the Design Council taking place across the United Kingdom. Dott 07 was based in the North-East of England with the help of its regional development agency, One NorthEast. It entailed a year of community projects, events and exhibitions, which explored what life in a sustainable region could be like – and how design could help us get there. The projects aimed to improve five aspects of daily life: movement, health, food, schools and energy. They were small but important real-life examples of sustainable, human-centred living which were designed to evolve and multiply in the years ahead. Their focus was on grassroots community projects, including more than seventy schools, local exhibitions and events in museums, galleries and rural sites. All events explored how design can improve our lives in meaningful ways.

The Dott 07 year culminated in a festival in Baltic Square on the banks of the River Tyne which showcased the results of the community projects and enabled all those involved to share their experiences. The projects were undertaken by community groups, facilitated by service design specialists from agencies such as Live|Work, Thinkpublic, Zest, staff from the Centre for Design Research at Northumbria in collaboration with Design Options. The projects included: Urban Farming, Low Carb Lane, MoveMe, OurNewSchool, New Work, Alzheimer100 and DaSH (Design and Sexual Health).

At the Dott festival closing, a presentation was made by Ezio Manzini, who suggested that the Dott 07 projects are a vital example of design with communities leading to the construction of community engagement, rather than design for constructing new products. He drew on his knowledge of the 'Creative Communities' of engagement EU programme undertaken by his own institution and other European Centres, which was also presented by his colleague Anna Meroni at ISDn3 (Meroni 2008). The main legacy of the Dott 07 projects was the transforming effect that engagement in the design process had on the lives and attitudes of community participants (Tan 2008).

Concurrent with the ISDn conferences and the Dott 07 programme, new doctoral projects were established at Northumbria University, Centre for Design Research, investigating different aspects of service design theory and practice, including:

1. An investigation of the design methods and process employed in and emerging from the Dott 07 community projects (Tan 2008)
2. The relationship between service touch-points and the way they are experienced and portrayed, which is now focusing on the design of services on the edge of society (edge services) that explore future service design theory and practice (Singleton 2008)

3. Investigating the capacity of transformation service design to move beyond social science methods in support of communities of craft practice in rural India

Other doctoral projects in service design methods and practice research have been completed and newly established at the Politecnico di Milano and at Exeter University (ISDn3 2008).

Multidisciplinary perspectives at the 'Designing for Services' workshop

The 'Designing for Services' workshop at the University of Oxford, SIAD Business School concluded in October 2007 with a series of presentations by researchers and academics from different disciplines (Kimbell and Seidel 2008). The author made a contribution linking perspectives of design method and theory with complex systems and service design practice. The main point of this presentation observed that design theory has adopted various 'centric' approaches to help resolve its relationship with design practice in the past: artifact-centred, human-centred, eco-centred, even author-centred (Young 2008b). However, experience-based design practice coupled with centric-specific approaches is inadequate for the current rate of contextual change. These methods may support the development of a conceptual idea, but they do not provide the tools and techniques required to analyse complex service ecologies, or to develop more human-centred solutions while dealing with multidisciplinary design situations (Hugentobler 2004).

In his paper, Young raised a question about the proclivity of the disciplines: 'Is there a natural lean of business to private sector contexts and that of design to public sector services' (Young 2008b). He pointed out that conspicuous consumption in service of product and artefact creation in the world of global business – as opposed to inconspicuous consumption in the context of service transformation in the world of public-sector services – polarises this debate. In Kimbell's concluding remarks (Kimbell and Seidel 2008) she observed that the representational tools and accounts of activities of service designers suggested to some participating academics that there was a reliance on binomial sets of relations grounded in traditional economic and manufacturing models of production and consumption. Yet Leonard's observation at Emergence 2007 was that there is a complex relationship between producing consumers also consuming producers – i.e. 'conducers' as well as 'prosumers'! This iterates the need for multiple perspectives, front-office and back-office consideration, and a composite theory and methodology to support practice.

A follow-on question raised by Young was: How, then, do we get rigour in service design practice? This question was prompted by a discussion with Jonathan Ive in July 2007 on the topic, where he maintained that service design as a field of practice is ill defined: unlike the rigour of good product design for manufacture, service design method and process has not yet been established and refined. Answering this question infers another: How do we make the shift from designers as executants of products to designers as executives of services? Hugentobler's prognosis is that this cannot be done without the introduction of systems thinking (Hugentobler 2004). More recent studies, e.g. Singleton (2008) and conferences, point to the need to consider the hybridisation of design methods and social science methods.

CIID Service Design symposium

This was an event held by the Copenhagen Institute for Interaction Design (CIID) in March 2008. The proposition of the symposium was that the service economy is the new paradigm in most knowledge-driven economies. Its purpose was to create awareness and impart practical knowledge, to provoke potential service design champions in relevant industries and foster new thinking in service innovation. It included speakers from the same range of service design groups and organisations as previous events but with a more Scandinavian flavour, mixed with keynotes by Bill Moggridge of Ideo and Ezio Manzini. Its programme covered understanding service design, academic explorations and industry case studies involving experimental practices. The symposium also sought a deeper understanding of how to harness design thinking as a strategy for best practices in the public sector.

David Chiu's sense of the CIID symposium was that, in contrast to previous service design conferences, where most presentations spent time introducing, explaining, defining, or making the case for service design, he found that the presentations had moved on to making the business case for service design, which seemed to fulfil Bill Hollins's plea that the success of a service should be measured financially (Chiu 2008). The symposium was not as effective in meeting its aim of delivering best-practice strategy for public-sector service development.

ISDN3, April 2008

This went on to investigate broader issues that contemporary designers face, with a special focus on how designers are addressing the complex situations that arise when designing with what John Thackara of Dott 07 calls 'real people' – as opposed to 'users' – in the design process. In this sense, ISDn3 had a similar but stronger emphasis on designing with people in contexts of public life than the CIID symposium. It did not follow the focus of previous service design conferences and events, which had been more concerned with what service design is and its relationship with business. The workshop was primarily for researchers in academia and the service design sector. Doctoral researchers were invited to share reflections on the *content* (topics and knowledge about designing with people), *processes* (research philosophies, methodologies and techniques) and *context* (situations and circumstances of designing). Its purpose was to maximise positive debate about the key design research issues arising when designing in this way. Specific questions at the workshop included:

1. What we can learn from public service explorations that include diverse communities of interest, e.g.: Dott 07 and Creative Communities?
2. Can service design methods connect inconspicuous consumption and people to policy?
3. Are 'edge services' particularly instructive?
4. Are there limits to the application of service design approaches and methods?
5. Can transformation service design methods go beyond the utility of methods from other disciplines to effect policy into practice, data into better qualities of experiences for real people in real-world contexts? If so, what kind of meta-theory could support this interdisciplinary work?

The value of ISDn3 was not so much in the outcomes of discussions that took place, but rather in the structuring of the discussions themselves. A scaffold for approaching the discussion within the ISDn3 workshop had been constructed from points derived from discussion of contemporary design research by Clive Dilnot (Dilnot 2008). He pointed out that traditional design research abstracts design thinking, creating a blind spot, which prevents the whole from becoming greater than the sum of its parts; that design naturally mediates relationships. Consequently, underlying questions to the discussion were:

1. How does theory inform practice?
2. How does informed practice create projects?
3. What are appropriate relationships to concepts of sustainability and human-centredness?
4. How does practice encourage research projects?
5. What theoretical constructs are being used to undertake doctoral studies in the arena of design for public services?
6. Does evidence of selected focal theory address Dilnot's observation of a contemporary paradigm shift from *quasi*-scientific to a complex dialogue with the social sciences?

Service Design Network Conference, Amsterdam, November 2008

This is the most recent conference on service design, which addressed an important thread of this review concerning representational tools and accounts of service design activity. In his reflection on the conference, Nicola Morelli (Morelli 2008) states that the presentations show that service designers have become proficient in involving users, stimulating participation and co-production. He observed that many presenters used similar representational tools and terminology, including the metaphor of the 'journey' to explain the service design process. His view was that the homogeneity of the methods and terminology is positive in that it perhaps shows that a tradition of working in this area was beginning to consolidate. However, he wonders whether the metaphor of the 'journey' is sufficient to describe the systemic perspective of service design. He reasons that journeys also use tools e.g. a train, a car or a flight; that the focus on the journey spotlights the experience but leaves the back office in the dark. When we travel we normally interact with other people, e.g. passengers and people who are supporting our journey, including people whom we never meet. Therefore, he also thinks that the perspectives of both the front and back offices are needed.

His conclusion was that the focus of the conference was on users, and this is opposite to service design perspectives in management and engineering books, which refer to services from the perspective of an organisation.

3 Drawing together the threads of the context, process and content issues

ISDn3 explained that what matters is debate around research projects which attempt the messy and complex business of negotiating practice and theory in the fields of designing *with* people. These fields are a matter of great interest and include subsets of complexity, systems, strategic, transformation and service design. Such designing needs to address aspects that Clive Dilnot referred to as the three horizons that define us – those of artifice *per se*, the market, and the emerging crises around social and environmental sustainability (Dilnot 2008). His point is: to think in relation to all three horizons concurrently is to embark on a critical project. It is to refuse to give the market absolute value, or any one 'centric' approach supremacy. Dilnot's proposal is that what design can do in relation to these horizons is to maintain the possibility of the future in forms other than a crisis situation!

Participants at service design conferences and events have tended at times to suffer from a 'gold rush' mentality where everyone is intent on staking their claim to titles, definitions, tools and methods within the field. The concern has been more about who owns or has originated this or that, whereas what we need to ask is how does theory work to inform practice and vice versa? How does theoretically informed practice work to create 'projects' which challenge the existing limits and models of understanding? Creating this relationship, mindful of the nature of the 'critical project', is critical to more sophisticated future practice and study of complex design problems, including service design.

Developing a more robust and sophisticated theoretical framework to support the complexity of service design practice requires us to embrace Dilnot's theoretical paradigm shift, resisting attempts to model design *quasi*-scientifically or technologically in favour of a much deeper and more complex dialogue or negotiation with the social sciences as a whole. Evidence that this is already happening emerged from ISDn3 both in terms of practical, social scientific methodologies, for example the use of ethnographic studies to support data collection, and analysis and rich representation as documentary video narration (Raijmakers 2006).

Dilnot again argues that this shift is natural because design is endemically the process through which the social is mediated in relation to products, services and artificial systems, and the concept of self. Establishing such a relation or set of relations will not be easy – especially at the theoretical level – because the pragmatics of practice often allow for more expedient adjustment of complementary ways of acting than entrenched theoretical models do. All this points to the fact that, at present, the design analytical community has almost no adequate 'institutional' means of exploring these relations – or of developing depth-expertise in design for social scientists or of the social sciences for designers – but things are starting to happen, and the service design and complexity science conferences and events have contributed significantly to this debate!

4 Conclusion

To conclude this review of the evolution of the theory–practice nexus for service design as an area of complex design practice I would like to focus on two important recurring aspects: first, the practical aspect of identifying appropriate representational tools to assist service designers; second, the theoretical aspect concerning the ontology of service design – what is the nature of service design and what theoretical construct best represents and informs the complexity of its practice.

Representational tools

Representational tools are needed the better to plan, act and present the design process and the embodiment of its values in the most realistic manner ('making real' – Buchanan 2008). A common observation of conferences and events on service design has been the use of the 'journey' as a metaphor to describe the service design process. This has its limitations (Morelli 2008). Perhaps a more realistic metaphor is the use of 'theatre'! This is not entirely new (Laurel 1993). This metaphor is being explored by Tan in her doctoral project (Tan 2008). Tan proposes that the front- and back-office activities that comprise a service are effectively represented by the actors, producers, directors, stage crew and support staff, the props, stage and backstage environments. The metaphor also concurs with the generally accepted importance of narrative to convey service value through the use of human stories, frequently referred to in the conferences and events. It also meets the aim of design research seeking to improve the design practitioner's (and other stakeholders') ability to navigate complex projects (Young 2008a), rather than it abstracting design and leading away from practice (Dilnot 2008).

Ontology of service design

The many-to-many interactions that comprise a service are hard to accommodate within the theoretical framework of contemporary design research and the conceptual frameworks familiar to most designers. There is a tendency to see a service as an immaterial good (where products are seen as material goods), or alternatively to ascend to a level of definitional abstraction that is poor at informing practice, as found in discussions of the constitution of services within economics and business, for example in Jean Gadrey's work (Gadrey and Gallouj 2002) and at Designing for Services and the CIID symposium. In the context of complex interacting systems, many-to-many equate well with more-is-more, and resonate with complexity science to generate an effective working and relational concept. This analogy supersedes the current tendency to refer to outmoded concepts from other disciplines – for example, to see services as 'co-produced' by clients and providers 'at the point of delivery', a definition that maintains traditional boundaries between consumption/client and production/producer, even though it concedes that they are problematic. This was the prevailing logic at ISDn2 and the Service Design Network Conference, which was questioned by Leonard at Emergence 2007 (Leonard 2008). The problem becomes quickly apparent if we look at the practices engaged in by a creative community as a service provided by some members of the community to others, e.g. Dott 07 (Thackara 2007), and EMUDE (Manzini and Jegou 2003) and discussed by Meroni at ISDn3 (Meroni 2008). Again, the analogy with complexity science is the notion of 'multi-agent systems', interacting with others in a neighbourhood using 'local' rules that emerge based on behaviour learned from interactions (Rzevski *et al.* 2006). This also takes us back to Jones's concept of a natural evolution, resulting in the same quality of rightness seen in organisms that have co-evolved (Jones 1977).

Singleton's doctoral study at Northumbria has co-explored theory and practice in the development of 'edge services'. He has come to see 'services' as performative socio-technical arrangements. This more general approach to what a service is, is informed by related work across a number of theories, particularly the 'assemblage theory' of Manuel DeLanda (DeLanda 2006) and his predecessors (Gilles Deleuze, Felix Guattari and Gilbert Simondon, etc.) and other interpreters of this philosophy (e.g. Brian Massumi), and the actor-network theories developed by Bruno Latour (Latour 2005) and others (Michel Callon and John Law). Singleton sees services as 'forces' in the socio-technical landscape, ways of moving around and bringing together people, things and messages, etc. As such, he has taken a view of services that sees them more as forms of 'soft infrastructure' that engender particular practices and modes of everyday life.

This approach has a number of advantages for service design. First, it obviates the distinction between the operation of services and the operation of communities, and also between services and the processes of production and consumption of physical products. Assemblage theories treat all of them in the same way, therefore allowing an integrated understanding of how a service shapes and is shaped by the urban context in which it is situated. Second, actor-network theory, an ontological standpoint on how the social is constituted borne originally from rigorous ethnographic processes, in particular pays attention to the role of non-human actors – the ways that physical objects, systems or ideas have 'agency' in their ability to mobilise and direct other (human and non-human) actors. Clearly a theoretical perspective that is also friendly to the practices of designers.

Singleton maintains that seeing a service as arrangements of people and things with interactions, more or less continually performed in order to be said to exist, has important critical implications. While designer commentators like Manzini *et al.* see bottom-up processes of organisation as something *a priori* desirable, the method of case-study selection that supports their conclusions implicitly favours services that fit effortlessly into the existing socio-technical landscape. But why should this be the case? From guerrilla gardening to online file-sharing systems, new forms of socio-technical arrangement are emerging that contest the legitimacy of familiar social actors – governments, corporations (Thackara 2005).

This review has demonstrated that complexity science (Johnson *et al.* 2007) in conjunction with social science theories (Dilnot 2008) provides an important way forward for design studies' consideration of theories and practices to contend with the complex problems found in service design contexts.

Acknowledgement

The author would like to thank Benedict Singleton and Lauren Tan (his doctoral candidates at the Centre for Design Research, Northumbria University) for their contributions to this review.

References

Buchanan, R. (2008), 'Emergence closing keynote', Pod-cast of presentation and debate, in *Emergence 2007*. Available at; http://www.design.cmu.edu/emergence/2007/, accessed 30 December 2008.

Chiu, D. (2008), available at: http://ciid.dk/service-design-symposium-recap, accessed 30 December 2008.

DeLanda, M. (2006), *A New Philosophy of Society: Assemblage Theory and Social Complexity*, London/New York: Continuum.

Dilnot, C. (2008), 'Criticality in design – the blind spot', available at: https://www.jiscmail.ac.uk/cgi-bin/webadmin?A2=ind0712&L=PHD-DESIGN&P=R2535&D=0&H=0&O=T&T=0, accessed 30 December 2008.

Dott 07, available at; http://www.dott07.com/go/what-is-dott, accessed 30 December 2008.

Gadrey, J. and Gallouj, F. (eds) (2002), *Productivity, Innovation and Knowledge in Services: New Economic and Socio-Economic Approaches*, Cheltenham: Edward Elgar.

Hugentobler, H. K. (2004), 'Designing a methods platform for design and design research', in J. Redmond, D. Durling and A. De Bono (eds), *Futureground*, Design Research Society, International Conference, Monash University.

ISDn, available at: http://www.cfdr.co.uk/isdn/programme.pdf, accessed 30 December 2008.

ISDn2, available at: http://www.northumbria.ac.uk/sd/academic/scd/whatson/news/listen/561569, accessed 30 December 2008.

ISDn3, available at: http://www.northumbria.ac.uk/sd/academic/scd/whatson/news/listen/808653?view=Standard&news=archive, accessed 30 December 2008.

Johnson, J., Alexiou, A. *et al.* (2007), 'Embracing complexity in design', in T. Inns (ed.), *Designing for the 21st Century*, London: Gower, pp. 129–49.

Jones, J. C. (1977), 'How my thoughts about design have changed during the years', *Design Methods and Theories*, 11 (1).

Jones, J. C. (1992), *Design Methods, Seeds of Human Futures*, New York: Van Nostrand Reinhold.

Kimbell, L. and Seidel, V. P. (eds) (2008), *Designing for Services: Multidisciplinary Perspectives*, Oxford: University of Oxford.

Latour, B. (2005), *Reassembling the Social: An Introduction to Actor-Network-Theory*, Oxford: Oxford University Press.

Laurel, B. (1993), *Computers as Theatre*, Reading, Mass.: Addison-Wesley.

Leonard, J. (2008), 'At your service: the blind men, the elephant, and the design of the world', Pod-cast of presentation and debate, in *Emergence 2007*, available at http://www.design.cmu.edu/emergence/2007/, accessed 30 December 2008.

Manzini, E. and Jegou, F. (2003), *Sustainable Everyday: Scenarios of Urban Life*, Milan: Edizioni Ambiente.

Meroni, A. (2008), 'Approaches to and reflections on sustainable social innovation and service design', available at: http://www.northumbria.ac.uk/sd/academic/scd/whatson/news/listen/808653?view=Standard&news=archive, accessed 30 December 2008.

Morelli, N. (2002), 'Designing product/service systems: a methodological exploration', *Design Issues*, 18 (3): 3–17.

Morelli, N. (2008), 'Service design conference in Amsterdam: spotlights and shadows', available at: http://nicomorelli.wordpress.com/2008/11/29/service-design-conference-in-amsterdam-spolights-and-shadows/, accessed 30 December 2008.

Raijmakers, S. W. J. J. (2006), 'Design documentaries: using documentary film to inspire design', PhD research thesis, Royal College of Art. Available at: http://www.stby.nl/stbydata/publications/phdthesi.pdf, accessed 30 December 2008.

Rzevski, G., Himoff, J. and Skobelev, P. (2006), 'Magenta technology: a family of multi-agent schedulers', in *Workshop on Software Agents in Information Systems and Industrial Applications* (SAISIA), Fraunhofer IITB.

Singleton, B. (2008), 'ARC-INC: an alternative view of what designing for sustainability might mean', available at: http://www.allemandi.com/cp/ctc/book.php?id=105, accessed 30 December 2008.

Tan, L. (2008), 'Design in public sector services: insights into the designs of the time 2007' (Dott 07), public design commission projects. Available at: http://www.allemandi.com/cp/ctc/book.php?id=110, accessed 30 December 2008.

Thackara, J. (2005), *In the Bubble: Designing in a Complex World*, Cambridge, Mass.: The MIT Press.

Thackara, J. (2007), *Wouldn't It Be Great if . . . ?*, ed. A. Perchard, London: Wardour, Design Council.

Young, R. (2008a), 'An integrated model of designing to aid the understanding of the complexity paradigm in design practice', *Futures*, 40 (6): 562–76.

Young, R. (2008b), 'A perspective on design theory and service design practice', in L. Kimbell and V. Siedel (eds), *Designing for Services: Multidisciplinary Perspectives*, Oxford: University of Oxford, pp. 43–4.

This page intentionally left blank

10

Metamorphosis of the Artificial

Designing the future through tentative links between complex systems science, second-order cybernetics and 4D design

Alec Robertson

1 Introduction

This chapter introduces the author's perspective in respect to his investigation into embracing the science of complex systems from an Art and Design background. It covers a wide context and approaches the creation of artefacts through the concept of '4D design'. The focus of the chapter is on the design of the relationships between 'everyday objects', people, and their environment involving a plethora of 'static' and 'dynamic' characteristics. Particular attention is paid to aspects of the performance arts by embracing 'movement' as an important element in the process of design and conceptualisation of consumer products and built environments.

The chapter starts with a brief outline of complexity theory to set the scene. This is followed by an outline of the notion '4D design' combined with a performance arts perspective – 'applied choreography' – and Second Order Cybernetics (C2), which has some resonance with both the notions of the 'science of complex systems' and 4D design. The notion of 'soft innovation' is introduced concerning non-functional product characteristics related to aesthetic appeal. The chapter concludes with some reflection and provocations as to future related possibilities for innovation in everyday products, services and systems in the built environment. This is intended to be a catalyst for further thought and modest inspiration.

The author acknowledges that the topics covered are rather *ad hoc*, and that traditional scientific approaches to discourse are not strictly followed – not least in respect of the fact that methods for research differ between disciplines, and that 'design enquiry' can be different from 'scientific enquiry'. There is less of providing rational systemic information on-a-plate with the hope that all recipients will get a homogenous message (this is touched upon later). Instead the chapter recognises the value of 'ambiguity', which can increase the chance of unique connections 'emerging' by individual perception.

2 Complex systems science (CSS)

'Complex systems' is a general term used to describe systems that are diverse and made up of multiple interdependent elements. Many agree that complexity can emerge from the interaction of autonomous agents – especially when agents are people (Bourgine and Johnson 2007). There are numerous reasons as to why a system might be considered complex, including having one or more of the following characteristics (Johnson 2007):

- many heterogeneous parts
- complicated transition laws
- unexpected or unpredictable emergence
- path-dependent dynamics
- network connectivities and multiple subsystem dependencies
- dynamics emerge from interactions of autonomous agents
- self-organisation into new structures and patterns of behaviour
- non-equilibrium and far-from-equilibrium dynamics
- adaptation to changing environments
- co-evolving subsystems
- ill-defined boundaries
- multilevel dynamics

There are numerous definitions of 'complexity', and any one of these characteristics can make systems appear complex (Johnson 2006b; Horgan 1995; Edmonds 1999). Generally the use of 'complexity science' methods involves managing or controlling a system of elements so that their interaction as a whole moves towards 'desirable' future paths or states, and away from undesirable ones. The 'complex systems science' as a whole encompasses notions that can be applied to both the design of the built environment and human situations.

Like 'complexity', the definition of 'design activity' is hard to pin down, as it encompasses multifarious activities and perspectives. At one extreme there is 'evidence-based design' (EBD) with a rigorous research element, and at the other there is 'creativity-based design' (CBD) which uses the rich ambiguity of talented artistic search. The definition of Herbert Simon (1969) for design is useful here as a basis.

> Design is the transformation of existing conditions into preferred ones.
> (Simon 1969: 55)

3 4D design

The notion of 'transformation', along with 'metamorphosis', is at the core of the concepts presented here, with the idea that there are designs designed that are not currently considered as being designs in a 'professional context', and it is timely to consider *new ways* for creating the artificial. A position taken is that one way can be characterised, in part if not in whole, by the definition of '4D design' along with ideas of cybernetics and complexity science. It is suggested that readers view the website www.4d-dynamics.net in conjunction with reading this section. Although the definition is close to the sciences of complex systems science and cybernetics, 4D design has a cultural and aesthetic context and a more intuitive way of looking at the artificial world involving people.

> A 4D Design is the dynamic form resulting from the design of the behaviour of artefacts and people in relation to each other and their environment.
>
> (Robertson 1995)

4D design focuses upon designing 'cultural expression' within dynamic situations of the everyday 'designed' world in the field of Art and Design along with 'utility'. The main characteristics of 4D design are depicted in Figure 10.1. This is a diagrammatic conception that crucially shows the relationship of the performance arts to functional actions of people and dynamic technologies fundamental to it. The diagram has four basic domains of knowledge; two cover the dynamics of intangible media and tangible artefacts – multimedia technology and robotics; the other two deal with the dynamics of people, first, within functional work – 'ergonomics' – and, second, 'play', focusing on the performance arts involving dynamic cultural expression and meaning. In Figure 10.1 there are also subset domains shown, and these are: the relatively new discipline of 'interaction design', focusing on the usefulness of digital technological objects; 'interface design', focusing on the usefulness of digital informational media, e.g. screens

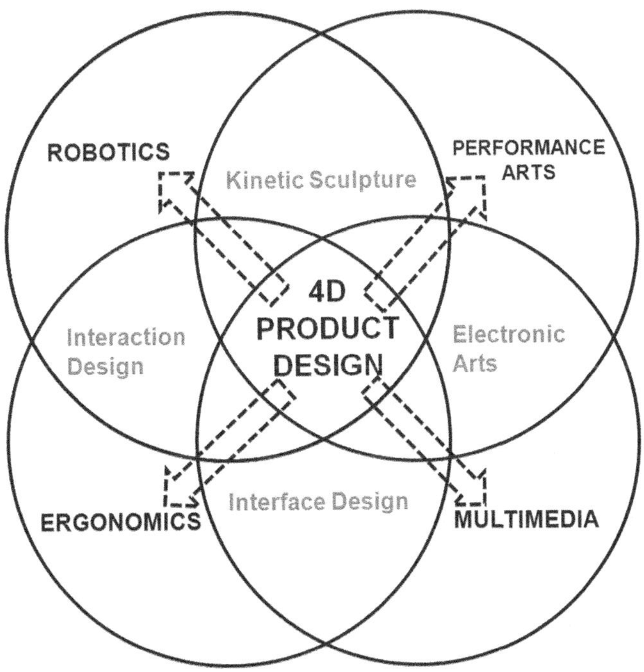

Figure 10.1 4D design diagram – Alec Robertson (1995)

and surfaces; the 'electronic arts', which deal with expression through intangible digital media including art installations; and 'kinetic sculpture', which focuses on dynamic expression of material art objects.

The arrows recently added through the '4D design' core of the diagram highlight that it can mainly involve the design of relationships between 'the artificial' in the form of digital multimedia and robotics technologies, or mainly involve 'people', encompassing both the utilitarian perspective of 'ergonomics' and the more playful 'performance arts'. In other words, 4D designs can result in artefacts alone acting in relationship to each other, such as robots dancing interactively with digital graphics on screens (although a human observer is assumed to be present), or mainly people acting in relationship to each other, such as the elegant performance of an up-market restaurant waiter with a customer (without much technology such as a portable credit card reader. The professional context of service design is a creative challenge (Robertson 1994).

The conception of 4D design here is limited to that of the 'everyday' contribution of Art and Design designers to the artificial world, where people are central, and where culturally rich dynamics are a main characteristic. It generally excludes systems where people are not present, such as the dynamics within an engine, and creations of the 'pure or fine arts' with no utilitarian purpose. In this context we can ask the question: What might the notion of 4D design with the science of complex systems contribute to artefacts and the actions of people in the 'everyday' built environment?

To help answer this question, the author designed and organised three symposia in close collaboration with the research cluster 'ECiD – Embracing Complexity in Design' (Johnson 2006b). (The author considers these events to be one type of 4D design.) Two events designed had the title 'More is More' (Robertson 2005, 2008). The first focused generally on the 'nature of design', and the latter on complexity in relation to the design of robotic devices with performative characteristics. The third event was called 'Magic in Complexity' (Robertson 2007) and focused upon multimedia game design from an arts perspective tentatively associated with 'complexity'. These events demonstrated that the field of Art and Design and the science of complex systems have shared interests to explore, with their complementary inherent differences, such as the use of different terminology and languages, personalities of 'different' people engaged, and idiosyncratic group behaviour that emerges from interdisciplinary work (Everitt and Robertson 2007).

The first symposium, 'More is More: Embracing Complexity in Design' (Robertson 2005), held on 16–17 December 2005, was the finale event of phase 1 of the 'Embracing Complexity in Design' research cluster; part of the UK AHRC and EPSRC research initiative 'Designing for the 21st Century'. The aim was to stimulate intellectual academic exchange, celebrate and disseminate the work of the ECiD cluster. It brought some members of the complexity science community together with several engaged in the Art and Design community. Figure 10.2 shows john chris jones giving the Symposium Dinner Address.

The second symposium, 'Magic in Complexity: Embracing the 4D Design-Arts' (Robertson 2007), took place on 23 February 2007. This event was part of the

Figure 10.2 'More is More' Event in the RCA Senior Common Room. Photo by Ismail Saray of ARTZONE, 2005

second phase of the 'Embracing Complexity in Design' research cluster – ECiD2. It comprised a symposium with keynote presentations, a networking soirée, and included Stimulus Talks and Serendipity Syndicate workshops, with interdisciplinary participation. The Serendipity Syndicates addressed five questions aimed to assist the ECiD2 research being done:

S1. How can the methods of complexity science assist digital games designers?
S2. How can design of 'play' in digital games inform research into complex systems?
S3. How can we create complex adaptive educational digital games?
S4. How can complexity help us to understand the enabling conditions of creativity and design?
S5. How can complexity theory be applied to the design arts in general?

Multimedia proceedings of this event are available via www.4d-dynamics.net

The five basic points below summarise the deliberations of the Serendipity Syndicates:

Summary Point 1: Understanding complexity theory helps designers to open up the games so that players can take part in the emergent design. Games design and complexity science is a fruitful area to explore further.
Summary Point 2: Play and Game design theory characteristics may be applied to Complexity theory characteristics.
Summary Point 3: Educational games can provide very simple visual solutions to very complex systems. This opens up the space for learners as they can appreciate small stimulates while grasping that they originate from a more complex whole.

Summary Point 4: In creative work involving complex systems, success is not readily measurable in the short term. This may be fundamentally at odds with the realities of complex systems.
Summary Point 5: Artistic work often explores how behaviours of systems emerge. Emergence can manifest itself both as part of the artistic process and as an outcome – look at artworks as systems rather than dissecting them into their constituent parts.

The third symposium, 'More is More 2: 4D Product Design for the Everyday' (Robertson 2008), had a basic tenet that new creative industries may well appear in the twenty-first century that we have yet to conceive of, and it is important to explore *avant garde* ideas for these related to 'complexity', robotics and the performance arts. This event included an afternoon symposium followed by an evening of public DIALOGUE in the Dana Centre of the Science Museum, London. '4D objects' were available to stimulate emergence of ideas for future design practice and research. Encouraging reflection by experts and the general public alike upon our relationship to 'dynamic objects' for 'real world' application as 'delightful' and 'useful' products, systems and services was the purpose of the day, along with dissemination of some interesting research being done. Figure 10.3 depicts a set of screen shots from its video proceedings to give a flavour of this event.

In summary, the events had a prime role of 'dissemination of research', with experimental multimedia online proceedings assisting this to a wide audience. The second event, 'Magic in Complexity', resulted in a special issue of a journal (Goodman *et al.* 2007); and the third, 'More is More 2', enabled dissemination of high-level design research directly to the general public, amongst other benefits. It was established that artists

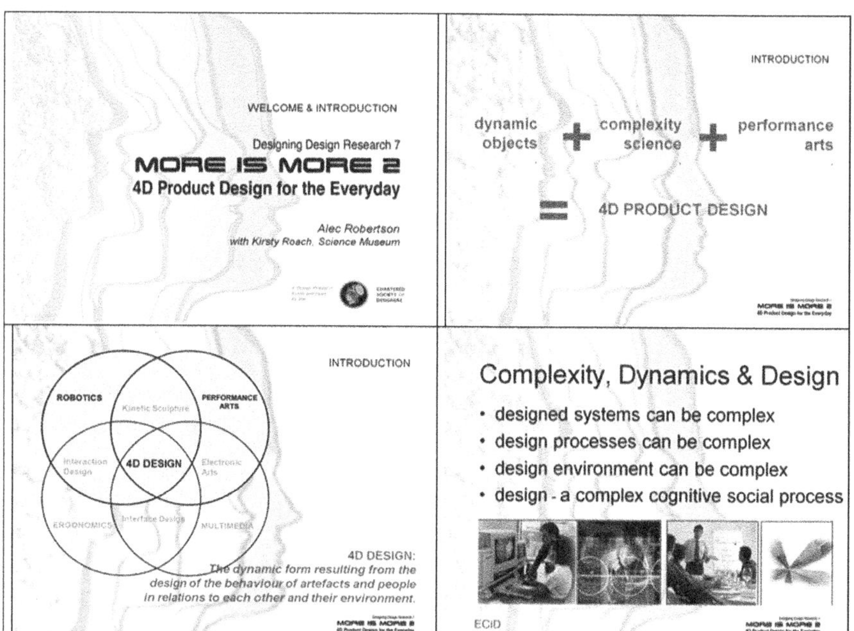

Figure 10.3(*a*) Sample video stills from 'More is More 2: 4D Product Design for the Everyday'

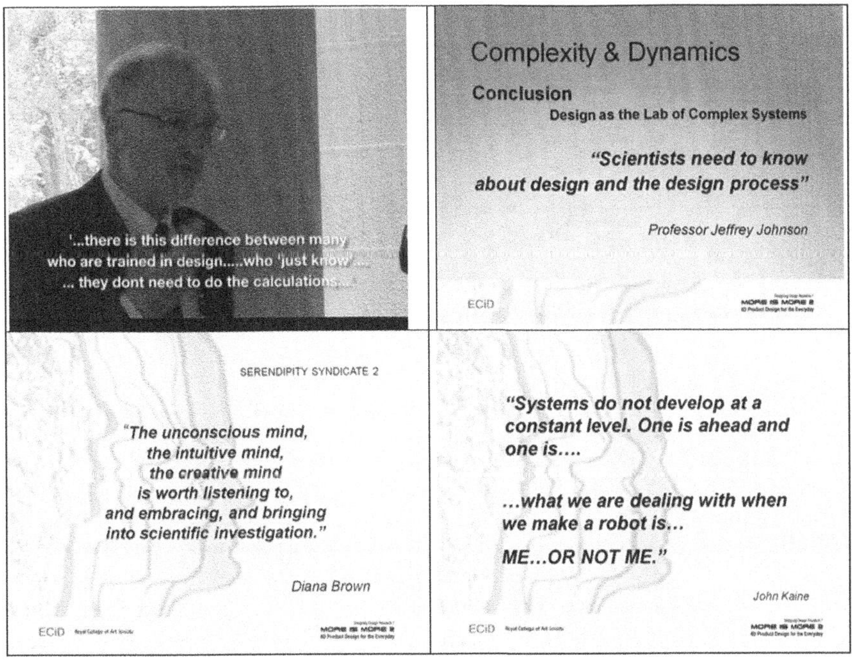

Figure 10.3(*b*)
Sample video stills from 'More is More 2: 4D Product Design for the Everyday'

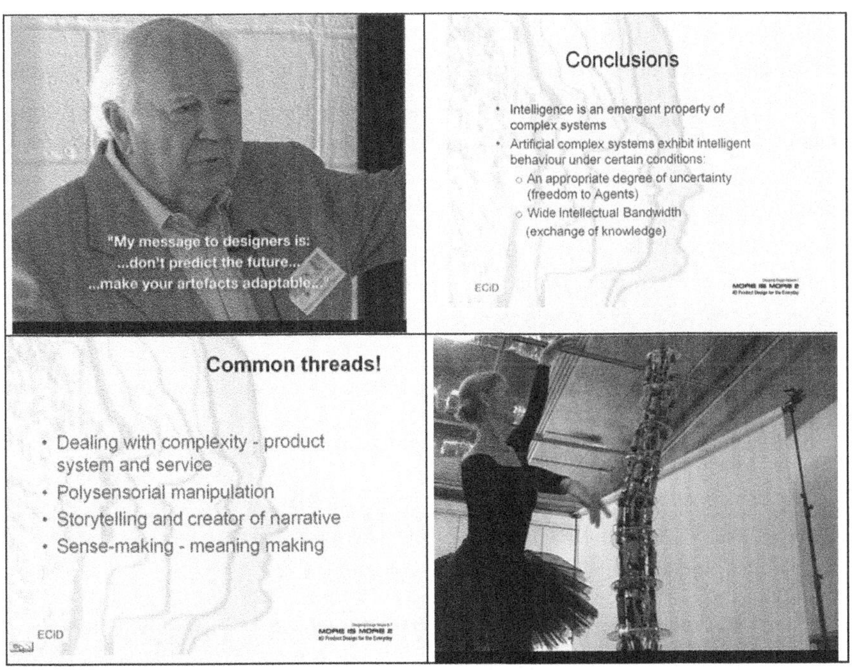

Figure 10.3(*c*)
Sample video stills from 'More is More 2: 4D Product Design for the Everyday'

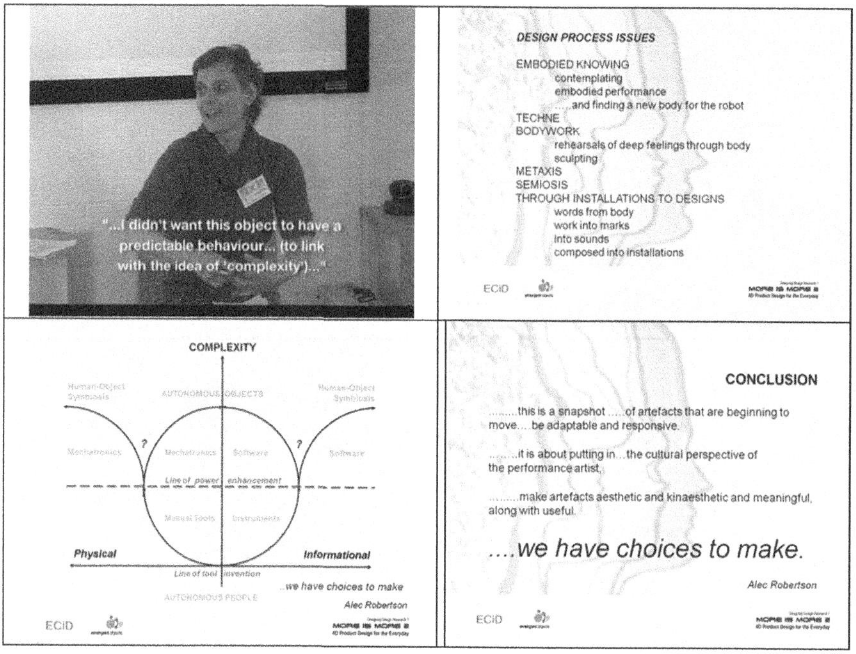

Figure 10.3(*d*) Sample video stills from 'More is More 2: 4D Product Design for the Everyday'

and designers can cooperate with engineers and complexity scientists and have shared interests. There was a recognition that designers are often capable of intuitively grasping 'complexity' when they design within their specialism, and Art and Design could offer much to this new frontier of 'complex artefacts' with its 'ways of visualisation' and 'ways of knowing'. Lessons were learned for organising future interdisciplinary events to maximise outcomes of such ventures, particularly for the effective 'capture of ideas' on the day. Participants on the day made connections on their own, and the author as both convenor and a participant has been able to make tentative connections generally in relation to his own viewpoint, and some are embodied in this chapter. Visitors to the On-line Proceedings at http://www.4d-dynamics.net/ddr7/ will, it is anticipated, make their own creative connections, too, as 'autonomous agents' in the spirit of the science of complex systems.

4 4D product design

So how can concepts of 4D design with complex systems science enable the creation of new kinds of 'artefacts' for the 'everyday – 4D products'? At the event 'More is More 2', 4D product design for the everyday is defined as:

4D Product Design = Dynamic objects + complexity science + performance arts

Robertson (2008)

Let us consider the notion of 'dynamic form' through the term of 'applied choreography'. This is a term proposed by the author as an attempt to encourage transdisciplinary work between the fields of 'design' and the 'performance arts' (Robertson and Woudhuysen 2001). The concept encourages application of useful choreographic knowledge to everyday life situations (outside theatre stages and without trained dancers). Sophia Lycouris has outlined a relevant theory in relation to architecture where 'space' is caused by the interrelationship between body, movement and space, and the act of design becomes the shaping not of buildings, but of space conceived in relation to a moving point of reference (in Robertson *et al.* 2007). She adds that interdisciplinary articulations have supported the development of conceptual frameworks for an understanding of architecture as a discipline which can accommodate change and instability, as well as material and conceptual flexibility (Brayer and Simonot 2002), and points out that various 'professional' architects, such as Lars Spuybroek and Peter Eisenman, have challenged the perception of architectural space, too. The advantage of creating architectural space with an integrated understanding of its dynamic potential, according to Lycouris, is that such space can increase the corporeal responses of the viewers or users, in the sense that as they move through the building they perceive space more intensely as a result of the generation of multiple physical sensations. The conclusion drawn is that expanding our understandings of choreography as a compositional method generally means that new design possibilities can arise. With the support of architectural theory that recognises the relevance of movement for users' experience of architectural space, such as that introduced by Sophia Lycouris above, new conceptual architectural possibilities can open up. An understanding of space as a dynamic entity (which includes objects and animate agents or people) presents a challenge to the static character of architectural design conceptions and manifestations generally.

4D design helps to make conceivable the integrated choreographic understanding of all manifestations of movement in a given physical space beyond 3D iconic form. In the architecture of public spaces, the concept of 4D design brings together physical objects, media and the activity of people within space, and can thus engender a dynamic multisensory expression of culture. With the tentative linking of choreography, architecture and complex systems through 4D design, and an emphasis on dynamics, interaction and relationships between the behaviour of artefacts and their users, it should be possible to expand the potential for collaborative ventures between disciplines. Some design research speculation on radical new design possibilities is left to later in this chapter.

5 C2 cybernetics

This brings us to the topic of 'cybernetics', which has some resonance with the science of complex systems and 4D design. A general view of this concept is given by Ashby (1956). More recently, the idea of second-order cybernetics, or C2, has been seen as increasingly pertinent. Ranulph Glanville highlights that an early cybernetics scholar, Heinz von Foerster, pointed out the absurdity of the traditional denial of 'the observer' in science, where there is a fictional creature through which knowledge is somehow immaculately generated (Glanville 2008). Foerster (1974) distinguished two types of cybernetics:

> First order cybernetics is the cybernetics of observed systems.
> Second order cybernetics is the cybernetics of observing systems.

Glanville depicts the difference between the two in the following way. 'First order cybernetics (C1) is concerned with circular systems, or systems of circular causality. In C2, we accept that the observer is "touched" (and touches) what goes on and there is circularity: the observed system is circular, but the observing system is also circular.'

To illustrate this, Glanville uses the classic example of a switch mounted on the wall to control a furnace which delivers heat to a room. He asks: 'What controls the switch, causing it to turn the furnace on and off?' Temperature is the answer provided, which in turn depends on the heat provided by the furnace. So it is the furnace that controls the switch, and each controls the other in fact. Glanville adds that the only reason why we call the switch the controller rather than the furnace derives from considerations of energy: the idea of a small amount of energy controlling a larger amount, which in turn entails a concept of 'amplification'. Cybernetics is primarily interested in flows of information rather than control; of physical energy, and the latter can be referred to as 'mechanisation' in the context of machines.

Another related C2 concept stressed by Glanville as important is 'conversation'. The view of communication based in Claude Shannon's 'Mathematical Theory of Communication' is that of 'passing coded messages accurately down channels with capacity to contain them' (Shannon 1948). Glanville highlights that, in C2, communication is not via coded messages that all people will receive as the same, but by conversational construction of pluralistic meanings that result. The sender may hope that all people receive a message well enough to get what s/he means as an aggregate. So, in C2, each communicator constructs their own meanings from the messages they pick up, which may well be different. In 'complexity science' terms, instead of considering one agent as being responsible for what happens, all agents control and maintain a system together as an aggregate of their individual presences, where each will perceive the situation differently.

For Glanville, 'conversation' is real 'interaction', and current use of the word 'interactivity' in computer and design fields is usually misconceived. So a challenge for designers is to consider the use of a product not as an action by a person on a machine or vice versa, but as a property of a system existing between them where each affects and is affected by the other. Importantly the system will involve 'movement' of some kind when observed externally: it will be 'dynamic'. In a circular system, in the sense of a cybernetic conversation, the '4D design' does not come from any one participant, but from the 'emergent property' of all participants acting together. It is shared and cannot be divided into the contribution of one or the other element, and each will perceive the situation differently.

The notions of 'interaction' and 'interface' design from the viewpoint of 4D design need expanding in this light. Paul Martin (1995) describes this idea of 'the space within' in the context of designing interiors of buildings from a 4D design perspective, where both the objects and the immaterial dynamics within the spaces are designs. Glanville calls this the 'inter-space', and adds: 'The challenge is to create this inter-space that will support conversational interaction' (Glanville 2008). Consideration of the participant in the 'inter-space' also involves use of a subject as its own subject (self-reference), as well as being one of multiple autonomous participants or 'agents'. C2 is a powerful concept and important for designing, especially where imagination and creativity are allowed to flourish. In a 'complex situation', 4D designing results in dynamic form both of the agents, and importantly, within 'the space between' them, through the creation of an unpredictable holistic 'emergent dynamic form' experienced by participants and observers. For example, imagine a person dancing with a robot where they are both in conversation through movement. Numerous luminous elastic bands are tied between them. Turn the lights off and you will see the dynamic form only 'in the space between'. The elastic bands are a metaphor for meaningful dynamic connections between them – a 4D design.

6 4D design innovation

Speculating about the possibilities of the conjunction between complexity science, 4D designing and C2 cybernetics for innovation requires brief consideration of changing ideas within the notion of 'innovation' itself. Definitions of concepts related to innovative activity, research and development (R&D) are found respectively in the Oslo and Frascati manuals (OECD 2002; OECD 2006). These are highlighted by Paul Stoneman, who adds to 'product innovation', 'process innovation' and 'marketing innovation' the concept of 'soft innovation' (Stoneman 2007).

> A soft innovation is defined as changes in either goods or services that primarily impact upon sensory perception and aesthetic rather than functional appeal . . . where a soft innovation may have different looks, touch, smell, aural patterns and will differently address personal preferences as to these.
>
> Stoneman (2007)

This definition concerns the value of 'aesthetics' in a wide range of industrial sectors from consumer products to architectural services outlined in Higgs *et al.* (2008). Aesthetic innovation is a subject of some research in terms of commerce for manufactured 3D product designs (see Marzal and Esparza 2007; Tether 2006). Paul Stoneman, with his business perspective, highlights that the measurement and judging of commercial aesthetic significance are poorly developed, although he adds that approaches that have been suggested might be 'influence upon others', 'the number of imitators', or the 'extent of copying'. At present, valuation of 'aesthetics' in the first instance is usually left to the intuition of experts who are acknowledged to be skilled judges in a field – not least, for example, Art and Design educated designers, and architects etc.

However, 'kinaesthetics' and other 'performative' qualities of 4D designs complicate assessment of aesthetic value in innovations. The perspective of 4D designing is a radical departure from the norm of 2D and 3D designing within the field of Art and Design, and it is a way of designing some sorts of activity that are not included in the categories of contemporary design education or 'professional design' (Robertson 1995). As a result, many such 'products' are not designed well. The 2D surfaces of many contemporary electronic consumer product interfaces, such as mobile phones, that largely comprise flat graphics are often too small to see or touch by many users. The 3D spatial design in many electronic goods is increasingly just to 'contain' the electronics as well, often in the form of anonymous 'black boxes', where possibilities for enabling 3D form to 'tell' users what to do through creating communicative 'affordances' (Krippendorff 2006) is there, but not exploited to full potential. This leads to the topic of dynamic 'affordances' possible with 4D design in consumer electronic products, which is, however, beyond the scope of this chapter.

So where are we going with all this understanding of 4D design with 'soft innovation', and its cousins C2 cybernetics and complex systems science?

7 Some speculation

The notion of 'metamorphosis' is at the core of possible innovation in 'the artificial'. At a basic level of complexity, 'modularity' is one physical way to encourage metamorphosis; as seen in the plug-in form of desktop computers with various levels of module sophistication, but also present in many other artefacts from Lego toys to buildings. With 'software' now embodied in many artefacts, including buildings, another way is by movement of the physical elements as articulated forms using performative robotics (McKinney *et al.* 2008). There is some speculation below on what innovation might result from using these notions to give more complicated levels of 'complexity' in 4D designs.

In the early 1990s there were notions of 'the intelligent building', cyber-control systems (in contrast to automatic), 'archionics engineering' (analogous to avionics for aircraft) and 'kinematic buildings' (in contrast to static) (Robertson 1993). It was advocated that a building could be as beautifully responsive as a plant when changing within its environment and, with gentle articulation of its components, as graceful as a ballet dancer. Later questions were posed like 'Is the 'automatic door' which opens as one approaches the beginning of buildings dancing with people? 'Can we look forward to buildings and a built environment that responds kinaesthetically with each other as well as efficiently, with subtle performances of buildings in our cities?' (Robertson

2007). Examples of such 'kinematic architecture', where a building incorporates motion through use of dynamic technologies, are beginning to appear (see Robertson 2008 i). This way the built environment may become more amenable and less controlling. Likewise there are indications that consumer products are becoming more responsive (see Robertson 2008 ii). With the 4D design perspective on dynamics, we can encourage dynamic architectural expression within the whole public experience; designs involving choreographic expression within articulated buildings, consumer products, media displays, traffic and people. What could be possible with notions of complex systems science, and specifically the interrelationships of 'autonomous agents' within cityscapes? Concepts of complex systems such as 'swarming' and flocking are useful. One impact may be on the 4D design of 'traffic' of people and vehicles.

Movement of commuters on pathways could create a kinaesthetic spectacle, as could traffic flow through urban roads. 'Intelligent' traffic signals embodying choreographic ideas could encourage vehicles to interact during acceleration and braking, creating delightful movement for participants and observers alike, as well as efficient traffic flow. On the motorways there is the phenomenon of 'herds' of vehicles, and if cars incorporated 'social' software they could communicate with each other with courtesy using smart materials and technologies, as well as adjusting their drive to increase safety for themselves. Imagine a particular make of car, such as an Audi, communicating with other Audi cars by colour changes, thus operating like the driver ritual of flashing indicators to respect good reciprocal driving behaviour. Similarly, robotic traffic lights could sense your car approaching and wave you though a junction if no car is waiting with a pleasant comment through the radio sound system.

In the high-street fashion boutiques, smart fabrics for *haute couture* clothes could flirt by being aware of the wearer's physiological responses through sensing their environment. Dynamic elements embedded in the fabrics could create a greeting display, and even be like the flamboyant peacock displaying his feathers (see Robertson 2008 iii). On the farms and in factories it may well be effective to have 'colonies' of machines (Rzevski 2008) working in the fields and production lines that move with performative qualities in a ritual dance of their own design while working.

We can ask what might be in a design 'research-exhibition' (Robertson 2006) of 4D designs today. First, it may well be more like going to the opera as it would be 'poly-sensorial', and 'performative', and possibly 'collaborative'. In addition it will involve 'circularity', 'sharing experiences', 'mutuality', 'reciprocity' (Glanville), 'adaptation' (Rzevski), 'semiosis' and 'habitus' (McKinney *et al.* 2008). Perhaps the above has enabled 'emergence' of ideas for such a research-exhibition or research-opera in you as a reader.

It could be asked: So what? Why do all this? The answer is the same as the answer to questions like 'Why have magnificent 3D architecture and interiors rather than everyone living in sheds' and 'Why have *haute couture* 3D garments and shoes, rather than all wearing uniform overalls and boots'. The issue is simply to transform existing conditions into preferred ones and create a 'delightful' artificial environment for living in.

References

Ashby, W. R. (1956), *An Introduction to Cybernetics*, London: Chapman & Hall.

Bourgine, P. and Johnson, J. (2007), *The Living Roadmap for Complex Systems*, EC ONCE-CS Report, http://complexsystems.lri.fr/main/tiki-index.php?page=living+roadmap. Accessed 9 September 2007.

Brayer, M. A. and Simonot, B. (eds) (2002), *Archilab Orleans 2002 conference proceedings*, 31 May–14 July, Orléans, France: Editions HYX.

Edmonds, B. (1999), 'Syntactic measuring of complexity', PhD thesis, University of Manchester, Department of Philosophy.

Everitt, D. and Robertson, A. (2007), 'Emergence and complexity: some observations and reflections on trans-disciplinary research involving performative contexts and new media', *International Journal of Performance Arts and Digital Media*, 3 (2): 239–52.

Foerster, H. von (1974), *Cybernetics of Cybernetics*, Urbana, Ill.: University of Illinois.

Glanville, R. (2008), Summary of Cybernetics Redux. Online discussion. On YASMIN Community List Email Post. http://www.media.uoa.gr/yasmin/

Goodman, L. *et al.* (2007), Special Issue, *Journal of Performance Art and Digital Media*, Bristol: Intellect Books.

Higgs, P., Cunningham, S. and Bakhshi, H. (2008), *Beyond the Creative Industries*, NESTA Technical Report, London, January.

Horgan, J. (1995), 'From complexity to perplexity', *Scientific American*, 272: 74–9.

Johnson, J. H. (2006b), 'Embracing complexity in design'. Principal Investigator, Designing for the 21st Century Initiative of AHRC/EPSRC projects. UK. http://www.complexityanddesign.net

Johnson, J. H. (2007), 'Embracing complexity in design', in T. Inns (ed.), *Designing for the 21st Century: Interdisciplinary Questions and Insights*, London: Gower, pp. 129–49.

Krippendorff, K. (2006), *The Semantic Turn*, Boca Raton, Fla: CRC Press.

McKinney, J., Wallis, M., Popat, S., Bryden, J. and Hogg, D. (2008), 'Embodied conversations: performance and the design of a robotic dancing partner', in *Proceedings of the 2008 Design Research Society Conference*, Sheffield, 16–19 July.

Martin, P. (1995), 'Spatial design beyond three-dimensional form' in A. Robertson (ed.), *4D Dynamics Conference Proceedings*, Leicester: De Montfort University, pp. 149–53. Also available at http://nelly.dmu.ac.uk/4dd/synd4h.html

Marzal, J. A. and Esparza, E. T. (2007), 'Innovation assessment in traditional industries: a proposal of aesthetic innovation indicators', *Scientometrics*, 72 (1): 33–57.

OECD (2002), *Proposed Standard Practice for Surveys of Measurement of Research and Experimental Development*, 3rd edn, Paris: DSTI, OECD.

OECD (2006), *The Measurement of Scientific and Technological Activities: Proposed Guidelines for Collecting and Interpreting Technological Innovation Data*, 3rd edn, Paris: Commission Eurostat.

Robertson, A. (1993), 'Speculation on the future of engineering the environment', in *Proceedings of the Environmental Engineering Conference*, 21 September. Leicester.

Robertson, A. (1994), '4D design: interaction of disciplines at a new design frontier', *DMI Journal*, Summer, pp. 26–30.

Robertson, A. (1995), '4D design: some concepts and complexities', in A. Robertson (ed.), *4D Dynamics: An International Interdisciplinary Conference on Design and Research Methodologies for Dynamic Form*, Leicester: De Montfort University, pp. 149–53. Also available at http://www.dmu.ac.uk/4dd//guest-ar.html. Accessed 5 July 2007.

Robertson, A. (ed.) (2005), *DDR5: More is More: Embracing Complexity in Design*. Event hosted by the RCA Society and ECiD1 D21C AHRB/EPSRC Project at Royal

College of Art, London, on 16 December. Web proceedings via http://www/4d_dynamics.net.

Robertson, A. (2006), '2D and paper to 3D and pixels: the research exhibition and its on-line media', in *EAD06 Conference*, Bremen: University of Arts, 29–31 March. http://ead06.hfk-bremen.de/. Also available at http://nelly.dmu.ac.uk/4dd/DDR5/ead06-paper.pdf.

Robertson, A. (2007), *DDR6: Magic in Complexity: Embracing the 4D Design Arts*. Event hosted by SMARTlab, University of East London and ECiD2 D21C AHRB/EPSRC Project on 23 February. Web proceedings via http://www.4d-dynamics.net

Robertson, A. (2008), *DDR7 More is More 2: 4D Product Design for the Everyday*. Event hosted by DANA Centre, Science Museum, London. Part ECiD1 D21C AHRC/EPSRC Project on 6 May. Web proceedings at http://www.4d-dynamics.net/ddr7/ Accessed 25 November 2008.

Robertson, A. (ed.) (2008), 4D-Dynamics Playlist. YouTube Playlist: Index "http://uk.youtube.com/profile_play_list?user=4ddynamics". Accessed 25 November 2008.

i. Playlist: Dynamic Architecture "http://uk.youtube.com/view_play_list?p=F1CE499D192E3B2E" Accessed 25 November 2008.
ii. Playlist: 4D Products. "http://uk.youtube.com/view_play_list?p=EBF84A4DDDB698E9" Accessed 25 November 2008.
iii. Playlist: Robotics. http://uk.youtube.com/view_play_list?p=12FCE83DF0ED8885 Accessed 25 November 2008.

Robertson, A., Lycouris, S., Johnson, J. H. (2007), 'An approach to the design of interactive environments with reference to choreography, architecture, the complex systems of 4D design', *International Journal of Performance Arts and Digital Media*, 2–3: 281–94.

Robertson, A. and Woudhuysen, J. (2001), '4D design: applied performance in the experience economy', *Body Space and Technology Journal (On-line)*, 1 (1). Available at http://people.brunel.ac.uk/bst/vol0101/index.html Accessed 5 July 2007.

Rzevski, G. (2008), 'Complexity and emergence in robotics systems design', in *DDR7 More is More 2: 4D Product Design for the Everyday*. Event hosted by DANA Centre, Science Museum, London. Part ECiD1 D21C AHRC/EPSRC Project on 6 May. Web proceedings at http://www.4d-dynamics.net/Talks-ddr7.html/ Accessed 25 November 2008.

Shannon, C. E. (1948), 'A mathematical theory of communication', *Bell System Technical Journal*, 27: 379–423, 623–56.

Simon, H. (1969), *The Sciences of the Artificial*, Cambridge, Mass.: The MIT Press.

Stoneman, P. (2007), *An Introduction to the Definition and Measurement of Soft Innovation*, Working Paper. Warwick Business School for NESTA.

Tether B. (2006), *Design in Innovation*, a report to the UK Department of Trade and Industry.

This page intentionally left blank

11

Embracing design in complexity

Jeffrey Johnson

1 Introduction

While designers willingly embrace the science of complex systems, most scientists rarely give design a second thought and thereby miss one of the most revolutionary aspects of the new science: *design, in the context of policy, is an essential part of the experimental method of the new science of complex systems.*

Currently few scientists today know anything about design as a process for understanding, creating and managing complex systems; but by the end of this century, if not by the end of this decade, design will be required study for complex systems science, alongside mathematics, statistics, computation and other core topics. Many of the systems that we find hard to understand are socio-technical systems – systems of systems – with tightly coupled physical and social subsystems. Most of these systems are *artificial*, meaning that they are in part or whole man-made – they are *designed* (Simon 1969).

In the old science it was possible to go into the laboratory, shut the door and exclude the universe outside from consideration. Magnetism, gravity and light appear to be decoupled from human affairs. This makes them relatively easy to study, and it is not surprising that science has triumphed in physics and chemistry. The science of complex systems is not like this since it is not possible to separate their social and physical subsystems and study them in isolation. More generally, the subsystems of complex systems cannot be studied in isolation:

> Science stands today on something of a divide. For two centuries it has been exploring systems that are either intrinsically simple or that are capable of being analysed into simple components. The fact that such a dogma as 'vary the factors one at a time' could be accepted for a century, shows that

> scientists were largely concerned in investigating such systems as *allowed* this method; for this method is often fundamentally impossible in the complex systems.
>
> (Ashby 1956)

Despite this, we do try to study subsystems as a way of understanding the complex whole. When we look at any system we seem to be drawn into recognising parts and classifying them, trying to understand their properties in isolation, and then trying to understand how they come together as a whole. As with any chicken-and-egg problem, the solution appears to be an iterative co-evolution between a story of the parts and a story of the whole. Design is a process that enables this way of thinking in the search for solutions to practical problems.

The difficulties of 'divide and rule' in complex systems science make it much more complicated than conventional science, with many enormous datasets interacting in computationally expensive ways. Performing experiments on complex systems is inherently expensive, and changing social systems requires moral and political authority. This puts great constraints on what experiments can be done and who can do them.

'Pure' science is the study of systems as they are, e.g. astronomy or the collection of biological specimens. In contrast the science of complex socio-technical systems is usually concerned with systems as they *ought* to be (Simon 1969), e.g. our bodies ought not to be diseased, the climate ought not to be changing as it is, the traffic ought not to be jammed, terrorists ought not be able to disrupt our lives, and the world economy ought not to be in crisis. The idea that systems ought or ought not to be in certain states or follow particular trajectories translates into that of *requirements*. It seems to be a 'law' of human systems that where there are requirements there will be human activity proactively trying to satisfy them, where this activity will involve a process defined as *design*.

2 What is design?

There are many models of the design process (e.g. Cross 1994; Pahl and Beitz 1996). In its simplest form (Figure 11.1), design is a human process that involves (1) recognising that there is a set of requirements for something and making them explicit, (2) formulating a *specification* for future systems as the properties they should have to satisfy the requirements, (3) generating more or less detailed descriptions, from sketches to blueprints, of possible future systems that meet the specification, (4) evaluating these descriptions as proposed solutions to satisfying the problem of meeting the specification and satisfying the requirements, and either (5) accepting a solution and implementing it, or (6) generating new potential solutions in the generate–evaluate *design cycle*, possibly finding that the problem is over-constrained (no solutions), or under-constrained (too many solutions), and as a consequence (7) changing the require-

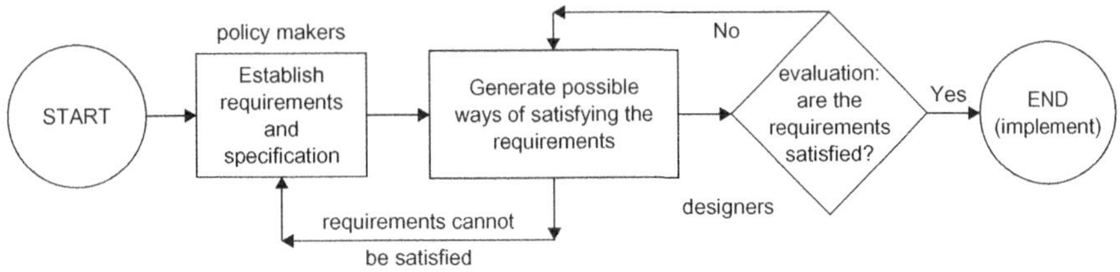

Figure 11.1 A simple model of the design process with two coupled feedback loops

ments and the specification. Thus *design can be considered to be a process to manage the co-evolution between problem formulation and solution generation.*

This process contrasts with problem-solving in the conventional sciences where the rules are that the problem cannot be changed. A student asked to solve the equation $y = e^{i\pi}$ would get little or no credit for suggesting that $y = x^2 - 2x + 1$ is a more appropriate problem, and that the solution is $y = 1$. But here it is explicitly asserted that the new science of complex systems inescapably involves changing the problem, and that embracing the design process is essential for progress in the science of complex systems. The *methodology* of complex systems science inescapably involves design.

3 What is prediction in complexity science?

Science can be viewed as the process of reconstructing dynamics of systems from data (Bourgine and Johnson 2006). Reconstructing the dynamics of systems requires at least (1) formulating a way of describing the instantaneous *state* of the system, and (2) formulating *transition rules* that transform the system to a new state at a new time.

In this context a *point prediction* is a statement that if the system is in a given state s_1 at time t_1, then it will be in state s_2 at time t_2. For example, physics provides point predictions of the positions of heavenly bodies. Making predictions and testing them by observation is a time-honoured way of testing theories – those that give incorrect predictions require revision.

There are many features that can make systems appear complex (Johnson 2006). Systems that have interactions between many autonomous agents generally have emergent dynamics that cannot be captured in a conventional way by sets of equations and require new methods for modelling their dynamics.

A large number of systems are *sensitive to initial conditions*, which means that small changes in the measurement of s_1 can result in large changes in the observed state s_2, so that point predictions become meaningless after some time horizon. For example, the weather can be predicted accurately over a few hours, but not over a few weeks. For such systems, the best one can do is estimate the probability that they will be in a given state at a given time. To test such a system, the best one can do is repeat experiments many times to see if the observed distribution is consistent with that predicted.

Many systems are *computationally irreducible*, meaning that future states cannot be given by equations and that each intermediate state must be calculated. For example, the cellular automaton called Conway's Game of Life is played on a grid of square cells. At time t each cell can have one of two states, alive or not-alive. The rules are: (1) non-live cell at time t will be alive at time $t + 1$ if exactly three neighbours are alive at time t, otherwise it remains non-live; (2) a live cell with four or more live neighbours at time t will be dead at time $t + 1$ (from over-population); (3) a living cell at time t with only one or no neighbours will be dead at time $t + 1$ (from under-population). It is easy to write a program to compute the next system state from the current state, and repeated applications allow future states to be computed. Marvellous configurations of dead and alive cells can emerge from these simple rules. However, to find the state at t_k starting at t_1 it is necessary to compute all the intermediate states – it is not possible to reduce this computation (as far as anyone knows).

In conventional science the transition rules are often given by differential equations that allow a particular outcome to be calculated for particular s_1 and s_2, whatever the value of t_2. This approach is computationally reducible, but cannot be applied to most socio-technical systems whose behaviour *emerges* from interactions.

Most complex socio-technical systems are both sensitive to initial conditions and computationally irreducible. In this case the best-known way to make predictions is to use *computer simulation* by programming the state representation and transition rules into computers, and applying the rules over and over again. Computer simulation suffers from what John Casti has called the 'can you trust it problem' (Johnson 2001), meaning that one can question computed predictions as possibly being flawed through a poor or incorrect model of the systems states and transition rules, inadequate or erroneous data, or errors in the program. None the less, computer simulation gives one of the best-known ways of mapping out the space of possible futures.

For our purposes, a *prediction* is a representation of the space of possible future states of a system from a given initial state, including where appropriate probability distributions across that space.

4 Experimentation in the science of complex systems: how can predictions be tested?

Predictions allow theories to be tested by experiment, and two types of experiment can be distinguished:

(1) *Non-interventionist observation*, e.g. Eddington's confirmation of Einstein's theory of the deflection of light by the sun's gravitational field by observing the shifted positions of stars during an eclipse, or Galileo observing the periodic motion of the cathedral chandelier. In both these examples the observers did not try to change the system they were observing.

(2) *Interventionist observation* intended to bring about a change, e.g. dropping a stone, changing the bank rate, inoculating a population, or building a bridge. In

these cases the observer takes action on the system under investigation with the explicit intention of bringing about change.

Most policy is effectively experiment, since the actions to be taken have not been tried before, the current state of the system has not been observed before, or both. Policy-makers initiate interventions in good faith in the belief that desired outcome will follow, but there is generally no conclusive scientific argument to support this.

Those responsible for policy make the second kind of experiment, deliberately taking action to make the system different. In this respect they can do experiments that scientists cannot, e.g. scientists cannot add extra lanes to a motorway or even change the traffic-light settings; they cannot change the bank rate or increase taxation; they cannot run the Olympic Games; and they cannot initiate a war or conclude a peace. Most of these projects require enormous amounts of money that scientists do not have, and many of them require an ethical and political authority that scientists do not have. *Only policy-makers have the mandate and the money to conduct really large-scale experiments on complex socio-technical systems.*

This means that scientists must either restrict themselves to non-interventionist observations, or they must find some way of persuading policy-makers to let them suggest interventions to test their predictions, including persuading the policy-makers to give them access to the data and their staff. The only way policy-makers will tolerate such disruption is if they believe the scientist can do something useful from a policy perspective, such as analysing the data in new and informative ways, or giving advice on the outcome of policy interventions that might lead to better outcomes. Sometimes scientists play this role as consultants and are part of the intervention-prediction team. Sometimes both scientists and policy-makers see mutual advantage in working together. And sometimes policy-makers can be wary, having previously worked with scientists who gave bad advice or wasted their time.

Most policy-makers dealing with complex systems have their own in-house scientists to advise them, so why is there need for other scientists to do experiments? The answer is that civil-servant scientists have different motivation and expectations from research scientists. While it is their job to use the best scientific knowledge available to solve practical problems, it is not their job to reflect on what they do and to create new knowledge (although many do). Once a project is finished, civil-service scientists may not have the luxury to document carefully the consequences of their interventions to test the underlying theory, because they will have been assigned to a new project and moved on.

If research scientists want to perform interventionalist experiments on large complex socio-technical systems, then they will have to align themselves with policy-makers. This is not easy; and, even if successful, the scientists will be junior partners in the enterprise because the policy-makers have the mandate and the money while the scientists have neither. Generally policy-makers will not accept scientists on to their team unless they are convinced that the scientists have something useful to offer throughout the collaboration.

5 Scientific predictions and the dynamics of policies

Prediction has been defined here to be a statement that when a system is in a particular state the model of the transition rules gives a distribution of possible states at future points in time. Suppose, then, that a scientist has formed a relation with a policy-maker, and that he has advised the policy-maker that given interventions will have various possible outcomes. In the simplest of worlds the policy-maker would implement the interventions, and the scientist would carefully record what happened and whether this corresponded to the theory. If it did, then the theory would be upheld. If not, the theory would have to be re-examined.

Unfortunately, it is not a perfect world. When asked what is most likely to blow governments off course, British prime minister Harold Macmillan replied: 'Events, dear boy, events.' In other words, implementation of government plans can be knocked of course by events that were not foreseen. To be useful to policy-makers, complex-systems scientists have to be able to understand such unexpected events.

Suppose a scientist gives the policy-maker the advice that a given intervention at t_1 will result in a range of outcomes at t_2. However, before t_2 some unforeseen event may occur knocking the system of trajectory. Frustratingly, the scientist will not be able to know what would have happened at t_2 and whether or not their prediction is validated.

It seems, then, that the best the scientist can do is make a sequence of predictions between t_1 and t_2 so that if an event knocks the system off trajectory at time $t_{1,2}$ with $t_1 < t_{1,2} < t_2$, then at least the prediction between t_1 and $t_{1,2}$ can be tested empirically. Assuming that the t_1 to $t_{1,2}$ approach were practical, the nature of testing the trajectory of the distributions could require new statistical theory.

But the scientist cannot stop work at $t_{1,2}$. There is a new scientific challenge: What were the meta-dynamics that knocked the system off trajectory at $t_{1,2}$? What can the policy-maker do next? This is illustrated in Figure 11.2 (Johnson 2009). Policy intervention is seen as giving the system a 'kick', trying to send it on a desired trajectory, while 'events' ranging from politics, money and even the weather keep knocking it off trajectory. The challenge is not just to reconstruct the dynamics of any particular kick, but also to reconstruct the *meta-dynamics* that result in the overall trajectory of what actually happens.

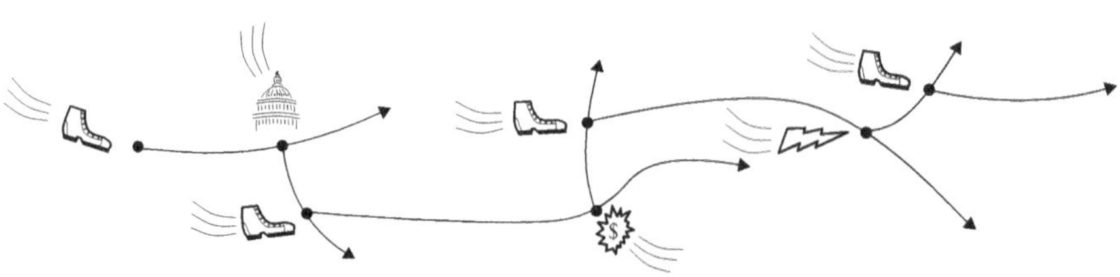

Figure 11.2 Policy is subject to many forces from many internal and external sources

6 Why complex-systems scientists must use the methods of design

Figure 11.3 extends the simple model of design in Figure 11.1 to the implementation of the design. In a perfect world the design plan will be implemented without problems, and the system will be built and function perfectly as specified by the designer. However, all kinds of events can knock this off course.

It is not uncommon for designs to be internally flawed with errors that only surface during implementation or construction. Anyone who has had building work done to their house will know that, even for simple systems, implementation can mean returning to the design phase. Of course most problems are minor and are easily overcome, but some can be serious and undermine the whole project. For example, the builders may unearth some unknown archaeological remains requiring the project be halted for a few months, when the weather will be worse, requiring the plan and possibly the whole design to be changed.

Many large projects have problems from time to time requiring minor or major revision to the design. Every deviation from the planned design means that the future behaviour of the system has to be tested, and this means that our scientist must be integrated into the design group in order to understand the problems when they occur and to be able to offer potential solutions. As such the scientist will be working intimately with the design team, going round the various design loops just like the designers and policy-makers. Even if they do not consider themselves to be designers, scientists will be deeply embedded in design practice and need to understand how it functions as a human system.

As we have presented it, the design and implementation of complex systems involves constantly making predictions within the generate–evaluate cycle and the design–implement cycle. Design becomes the *method* of making and testing predictions about the system before it exists, while it is being implemented, and while it

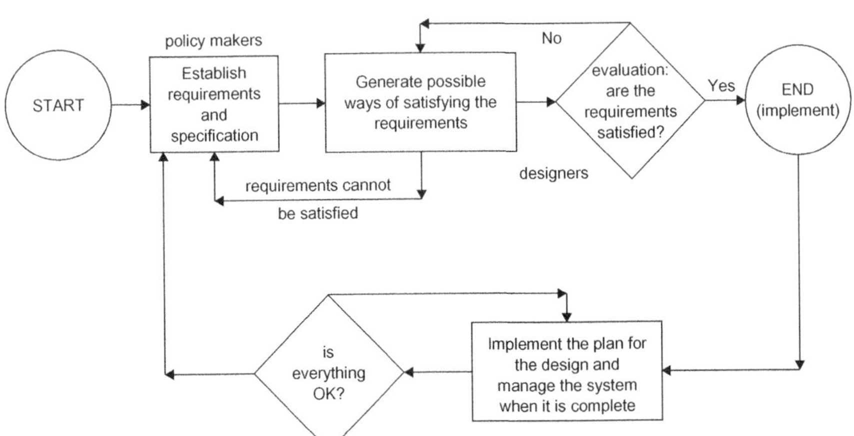

Figure 11.3 Design may co-evolve with its implementation

is being used. *Scientists wanting to test their predictions of complex socio-technical systems will have to embrace design.*

7 Illustrative examples

7.1 Pedestrian movement in public places

The abstract of a paper (Helbing *et al.* 2007) on the dynamics of crowd disasters reads:

> Many observations in the dynamics of pedestrian crowds, including various self-organization phenomena, have been successfully described by simple many-particle models. For ethical reasons, however, there is a serious lack of experimental data regarding crowd panic. Therefore, we have analyzed video recordings of the crowd disaster in Mina/Makkah during the Hajj in 2006. They reveal two subsequent, sudden transitions from laminar to stop-and-go and 'turbulent' flows, which question many previous simulation models. While the transition from laminar to stop-and-go flows supports a recent model of bottleneck flows, the subsequent transition to turbulent flow is not yet well understood. It is responsible for sudden eruptions of pressure release comparable to earthquakes, which cause sudden displacements and the falling and trampling of people. The insights of this study into the reasons for critical crowd conditions are important for the organization of safer mass events. In particular, they allow one to understand where and when crowd accidents tend to occur. They have also led to organizational changes, which have ensured a safe Hajj in 2007.

This research illustrates many of the points made in this paper. The Hajj is the annual pilgrimage of about 3 million people to Mecca, and it is clear that for policy-makers this event *ought* to be safe. The paper is co-authored by two scientists and an official from the Saudi Arabian Ministry of Municipal and Rural Affairs, illustrating what we have claimed is an essential cooperation between scientists and policy-makers. The paper itself has many technical details illustrating the availability of data to the scientists, and it concludes: 'Based on our insights in the reasons for the accidents during the Hajj in 2006, we have recommended many improvements,' showing that in this case the scientists were indeed acting as designers.

7.2 Complexity science in business

Rzevski (2009) gives the following examples of successfully developed and implemented complexity management systems: (1) managing in realtime a fleet of 2,000 taxis for a transportation company in London; (2) managing in realtime a large fleet of car rentals for one of the largest car rental operators in Europe; (3) managing in realtime 10 per cent of the world capacity of crude-oil sea-going tankers for a tanker management company in London ; (4) resolving clashes in aircraft wing design for the largest commercial airliner in Europe; (5) realtime scheduling of a large fleet of trucks transporting parcels across the United Kingdom; (6) agent-based simulator for modelling the airport and in-flight, RFID-based, catering supply chain, luggage-handling processes, and passenger processing for a research consortium in Germany; (7) selecting relevant abstracts for a research team using agent-based semantic search for a genome-mapping laboratory in the United States; (8) discovering rules and patterns in data using agent-based dynamic data-mining technology for a logistics company in the United Kingdom; (9) managing social benefits for citizens supplied with electronic ID cards for a large region in Russia.

Since these are all commercial applications of complex systems science, it is clear to see that in each case the system *ought* to have specified properties. In this case Rzevski worked closely with the policy-makers in the companies that commissioned this work. In all cases he was involved in the design process that resulted in these successful applications.

7.3 Epidemiology

A paper by Barrett *et al.* (2005) presents the EpiSims simulation system which is used to study how social networks spread disease. The authors write:

> Public health officials have to make choices that could mean life or death for thousands, even millions, of people, as well as massive economic and social disruption. And history offers them only a rough guide. Methods that eradicated smallpox in African villages in the 1970s, for example, might not be the most effective tactics against smallpox released in a U.S. city in the 21st century. To identify the best responses under a variety of conditions in advance of disasters, health officials need a laboratory where 'what if' scenarios can be tested as realistically as possible. That is why our group at Los Alamos National Laboratory (LANL) set out to build EpiSims, the largest individual-based epidemiology simulation model ever created.

This paper again illustrates scientific research that is closely aligned to policy and designing procedures to manage epidemics should they occur.

Elsewhere Barrett has proposed the idea of an 'Informatic Description of Causes' that may open up potentially new avenues for science (Johnson 2007). His idea is that simulation provides a way of investigating causes or is a way – possibly the only way – of projecting complex systems into the future. If the simulation used in this example of epidemiology were successfully used to control an epidemic, the prediction would be that the given intervention results in no epidemic, but this is impossible to test because the experiment cannot be repeated without the intervention in the expectation of observing an epidemic.

8 Conclusion: design in the science of complex systems

It has been argued that design is an essential part of complex systems methodology and *in-vivo* experimentation, and that, to make progress, complexity science must embrace design. To understand the dynamics and meta-dynamics of complex systems, the scientist must be embedded in real projects, working alongside policy-makers, who are the only ones with the mandate and the money to make purposeful interventions intended to change the system trajectories to what they *ought* to be.

It has been shown that some complex-systems scientists are already deeply engaged in the design process, collaborating with policy-makers over significant periods of time. In all these cases the scientists provide policy-makers in the public and private sectors with useful theories that enable the future behaviour of the system to be predicted. It is clear that these scientists could not have done their research isolated in the laboratory and detached from policy and implementation of policy through design.

It has been shown that design is a process of co-evolution between problem and solution. Design is a way of discovering the problem at the same time as solving it, and this is a new way of thinking for many scientists. The iterative generate–evaluate process of design is a spiral rather than a cycle, because on every iteration the designer learns. The design process discovers systems that are adapted to the environment at any time. As the environment changes, the fitness of any proposed design solution may change, and designs that were once well adapted may become less well adapted, while new and innovative design may be enabled.

The predictions that are useful are system-specific. Design projects and their implementations take many twists and turns. Whether or not a given action for intervention will result in a given outcome may be very important at one moment but not at all important at another. Thus the predictions that have to be made during the design process are shaped by it. There does not appear to be a set of Platonic design problems that can be sequentially solved to produce eventually a computationally reducible design theory, and this is a strength of design. Design adapts to the socio-technical environment. Design is able to respond to completely new and unknown problems. It can create completely new kinds of artificial systems with no *a priori* science, and the inventor-designer is the first scientist to explore the new universe that has been created.

Although we speak of 'the designer', complex socio-technical systems are designed and managed by teams of people with interacting specialisms. Scientists able to predict the dynamics and meta-dynamics of complex systems are essential members of these teams. This gives research scientists the opportunity to engage with policy-makers in the policy-directed *in-vivo* experiments of designing, implementing and managing complex systems. It is only by embracing design that research scientists can ensure that the experiments are instrumented in ways that allow science to progress through *in-vivo* scientific experimentation.

References

Ashby, W. R. (1956), *An Introduction to Cybernetics*, London: Chapman & Hall.

Barrett, C., Eubank, S. and Smith, J. (2005), 'If smallpox strikes Portland', *Scientific American*, March: 42.

Bourgine, P. and Johnson, J. (2006), 'The living roadmap for complex systems science', Ecole Polytechnique, Paris. (http://css.csregistry.org/tiki-index.php?page=Living+Roadmap&bl=y)

Cross, N. (1994), *Engineering Design Methods*, Chichester: John Wiley.

Helbing, D., Johansson, A. and Al-Abideen, H. Z. (2007), The dynamics of crowd disasters: an empirical study', *Physical Review* E 75, 046109. DOI: 10.1103/PhysRevE.75.046109 arXiv:physics/0701203v2, http://arxiv.org/PS_cache/physics/pdf/0701/0701203v2.pdf

Johnson, J. H. (2001), 'The "Can you trust it?" problem of simulation science in the design of socio-technical systems', *Complexity*, 6 (2): 34–40.

Johnson, J. H. (2006), 'Can complexity help us better understand risk?', *Risk Management*, 8 (4): 221–6.

Johnson, J. H. (2007), 'Complex systems for socially intelligent ICT: science, policy and implementation', in J. Johnson, M. Kirkilionis and J. Fernandez-Villacanas (eds), *Complex Systems Forum, ECCS'07*, Milton Keynes: The Open University.

Johnson, J. H. (2008), 'Science and policy in designing complex futures', *Futures*, 40: 520–36.

Johnson, J. H. (2009), 'Policy, design and management: the in-vivo laboratory for the science of complex socio-technical systems', *Proc Complex '09*, 23–5 January 2009, Shanghai.

Pahl, G. and Beitz, W. (1996), *Engineering Design: A Systematic Approach*, London: Springer.

Rzevski, G. (2009), 'Application of complexity science concepts and tools in business: successful case studies, mimeo (on the Complex Systems Society website).

Simon, H. (1969), *The Sciences of the Artificial*, Cambridge, Mass: MIT Press.

This page intentionally left blank

Index

Page numbers in *Italics* represent tables.
Page numbers in **Bold** represent figures.

For Product Safety Concerns and Information please contact our EU representative GPSR@taylorandfrancis.com Taylor & Francis Verlag GmbH, Kaufingerstraße 24, 80331 München, Germany

Printed and bound by CPI Group (UK) Ltd, Croydon, CR0 4YY
02/07/2025
01909228-0016